ELEMENTARY QUANTUM MECHANICS

ELEMENTARY QUANTUM MECHANICS

BY

R. W. GURNEY, M.A., PH.D.
Research Associate in the University of Bristol

CAMBRIDGE
AT THE UNIVERSITY PRESS
1934

CAMBRIDGE
UNIVERSITY PRESS

University Printing House, Cambridge CB2 8BS, United Kingdom

Cambridge University Press is part of the University of Cambridge.

It furthers the University's mission by disseminating knowledge in the pursuit of education, learning and research at the highest international levels of excellence.

www.cambridge.org
Information on this title: www.cambridge.org/9781107586352

First published 1934
First paperback edition 2015

A catalogue record for this publication is available from the British Library

ISBN 978-1-107-58635-2 Paperback

CONTENTS

PREFACE

The inclusion of sixty-seven diagrams in this book has enabled me to treat the problems of quantum mechanics by graphical methods. At the same time I have tried to present the principles in a form congenial to the experimentalist, keeping always in view the physical significance and actual numerical magnitudes of the various quantities. Two chapters have been devoted to the problems of valency and the properties of molecules. I have not discussed in any detail the scattering of electrons or the use of wave-packets, which have been treated in an elementary way in other text-books, for example in Mott's *Outline of Wave Mechanics*.

My thanks are due to the editors of the *Physical Review* and to Professor H. E. Whyte for the loan of the block from which fig. 10 is printed.

I wish to express my gratitude to Professor N. F. Mott both for criticising the book in manuscript and for assistance in proof reading.

R. W. G.

BRISTOL, 1934

CHAPTER I

§ 1. Nowadays when a physicist wishes to investigate a molecular or atomic problem, often the first thing he does is to draw an energy diagram. He draws a graph, plotting the potential energy of the system as ordinate against some co-ordinate as abscissa. For example, it may be the energy of mutual attraction or repulsion of two atoms plotted against their distance apart, or it may be the potential energy of an electron in a molecular electrostatic field. This way of tackling problems is characteristic of quantum mechanics, and is due to the fact that the potential energy V of the system, as a function of the co-ordinates, occurs in every form of the Schroedinger wave equation.

Facility in the rapid construction and rapid interpretation of these energy diagrams is easily acquired, and is of great help in understanding the innovations of quantum mechanics. We will therefore first consider energy diagrams according to classical laws, beginning with the simplest problem. The reader who is already familiar with these graphical methods may prefer to pass on to § 2.

If we throw a mass m vertically upwards with an initial velocity v, it will rise to a height equal to $v^2/2g$, where g is the acceleration due to gravity. During its flight, the total energy of the body, W, neglecting air-resistance, will be constant. If then we plot W as ordinate against h as abscissa, we obtain a horizontal straight line, such as AQ, fig. 1. As the body rises, it acquires potential energy V, at any height h equal to mgh. Plotting this on the same diagram, we obtain a straight line BD, whose slope is mg. The vertical distance at any point between the two lines, such as AB, gives of course the kinetic energy of the body at the corresponding height. As the body rises, the potential

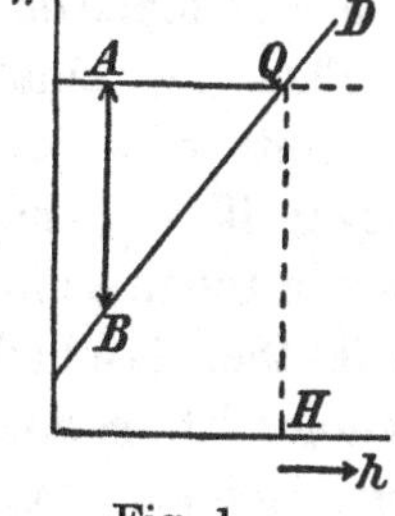

Fig. 1

energy goes on increasing until it becomes equal to the total energy W, when the two lines intersect; the kinetic energy has obviously fallen to zero, i.e. the particle has come to rest at a certain height H corresponding to the vertical dotted line through Q. And according to classical mechanics the particle will return from this point. The suggestion that the particle might rise to a height beyond H is absurd, because its potential energy would be greater than its total energy, and its kinetic energy would need to be negative, which is meaningless. The region beyond H is therefore a forbidden region into which particles of energy W cannot enter.

The process of retardation and stoppage, illustrated in fig. 1, is at the basis of the stability of the material universe. When in the nineteenth century the idea was developed by Clausius and Maxwell that all heat was due to the unceasing rapid motion of molecules, there was no difficulty as regards gases; on the contrary, the hypothesis accounted for their fluid properties very well. But for solids the position was different. To quote from the *Encyclopædia Britannica*: The distances traversed by the atoms of a solid are very small in extent, as is shown by innumerable facts of everyday observation. For instance, the surface of a finely carved metal (such as a plate used for steel engraving) will retain its exact shape for centuries, and again, when a metal body is coated with gold-leaf, the atoms of the gold remain on its surface indefinitely; if they moved through any but the smallest distances, they would soon become mixed with the atoms of the base metal and diffused through its interior. Thus the atoms of a solid can make only small excursions about their mean positions.

Thus the kinetic theory of heat was reconciled with the stability of solid matter by using the idea of allowed regions where $W > V$, and forbidden regions with a sharp boundary between them, as in fig. 1. It is supposed that for each atom of a solid there is a little allowed region in which the atom moves, the whole of the rest of space being the forbidden region. In whatever direction the atom moves, its potential energy rapidly increases, its kinetic energy meanwhile decreasing, as in fig. 1, until the atom is brought to rest and turned back again at the surface where $W = V$. The potential energy along any straight line will be of the form of fig. 2, so that the atom is confined in a kind of potential bowl or box. In this respect the atom resembles the bob of a pendulum, whose potential energy plotted against the displacement will always be like fig. 2, the kinetic energy being greatest in the middle. The width of the allowed region for the atom depends upon the total energy which it has, just as the amplitude of oscillation of a pendulum or of a tuning-fork depends upon its energy. Thus any line PQ lying above AB represents a higher total energy and possesses a wider allowed region. The V-curve may be symmetrical as in fig. 2, or else, when the retardation is much more rapid in one direction than in the other, it may be very unsymmetrical, as in fig. 49 on p. 88.

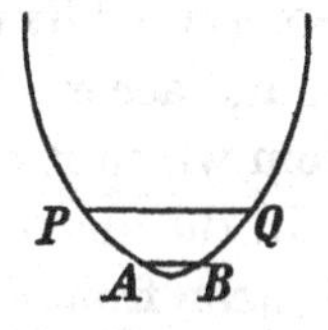

Fig. 2

It is not only in a solid that atoms must be pictured as confined to a small region as in fig. 2, but the same is true for any atom which forms part of a diatomic or polyatomic molecule. Each atom remains attached to the others, because it is in a potential box, surrounded on all sides by a forbidden region. Or, rather, it remains attached so long as its total energy W is less than V on all sides. If at any time the atom acquires so much energy that $(W - V)$ is greater than the depth of the potential bowl, fig. 49, the atom can escape, i.e. the molecule can dissociate.

The next step will be to discuss the motion of an electrically charged particle in an electrostatic field. Let A and B in fig. 3a be two parallel metal plates charged respectively positive and negative. And suppose we throw a ray of ultra-violet light on to the middle of A, so that a photoelectron is ejected from the metal. We will suppose that it is ejected at right angles to the surface, as indicated by the arrow. When the electron has escaped from the metal it moves against a uniform retarding field, in which its potential energy must be represented by a straight line, such as DF in fig. 3b. Now the particle may, or may not, have sufficient energy to get across to the opposite plate. If the total energy of the particle is such as to be represented by the line MN it will arrive; in fact, it will reach the plate with the small kinetic energy NF. If, on the other hand, the total energy is only such as to be represented by the line GC, then the electron will come to rest at Q, and according to classical mechanics will return to the plate from which it came.

In the above we have not considered the initial escape of the electron from the interior of the metal. This will be very similar to the familiar escape of a molecule from the surface of a liquid, a case which we shall treat first. Suppose a straight line AB be drawn perpendicular to the surface of a homogeneous liquid, as

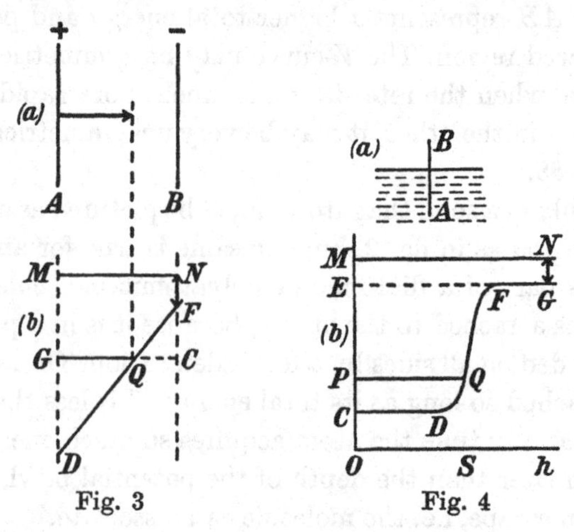

Fig. 3 Fig. 4

in fig. 4a; and consider what will be the potential energy of a molecule along this line. In fig. 4b let the abscissae be the height h along the line AB, and let S represent the point where AB cuts the surface of the liquid, the liquid being to the left of S and the vapour to the right. A molecule at rest in the vapour possesses more energy than a molecule at rest within the liquid; let this difference in potential energy be represented by CE, so that FG represents the potential energy outside the surface, and CD that within. Then the potential energy of a molecule along the line AB will be given by a curve like $CDFG$.

We know that when a molecule approaches a liquid surface from within with an energy greater than a certain critical energy, it can escape from the liquid. This case will be represented by a molecule moving along the line AB with kinetic energy equal to CM. It will be heavily retarded as it approaches the surface, but will escape with the small residual kinetic energy NG. It is of course only very exceptional molecules which approach with kinetic energies as great as CM. The average kinetic energy will be much smaller; let it be equal to CP. For this molecule the region beyond Q is in classical mechanics a forbidden region, so that the molecule is turned back from the boundary at Q. If P and M have been chosen so that CP and GN are the *average* values of the kinetic energy in the liquid and vapour respectively, then the quantity represented by the vertical distance PM is just the latent heat of vaporisation of the liquid at the temperature considered.

We may now return to the problem of the extraction of an electron from a metal. At room temperature the spontaneous escape of an electron from a metal surface is a comparatively rare occurrence, since the latent heat of evaporation of metallic electrons is many times larger than for molecules of liquids, such as water. But the energy diagram of fig. 4 may be used if we suppose the scale of ordinates to have been changed. Let AB in fig. 4a be a line drawn perpendicular to a metal surface, and let S in fig. 4b now represent the point where AB cuts the surface of the metal, a vacuum being to the right of S, and metal to the left. By considering an electron at rest inside the metal and an electron at rest in the vacuum, we obtain the potential energy curve $CDFG$. An electron having a kinetic energy such as CP is turned back at the surface. If, however, light is incident on the surface, and the electron absorbs from the light a quantum of energy equal to PM in fig. 4b, then the electron may escape from the metal with a residual kinetic energy equal to GN; this is the photoelectric effect.

If, instead of considering only one surface of a metal, we consider the potential energy along a line drawn right through a piece of metal, we shall have the complete curve of fig. 5, showing how an electron is turned back at either surface. And this diagram would apply equally to a molecule in a drop of liquid. In either case any horizontal line which cuts the V-curve (such as W in fig. 5) represents the energy of a particle which is confined to the "potential box", while any W lying wholly above the V-curve will be the energy of a free escaping or entering particle.

In figs. 4 and 5 we have been dealing with large-scale systems; the allowed region was the whole volume of the drop of liquid or

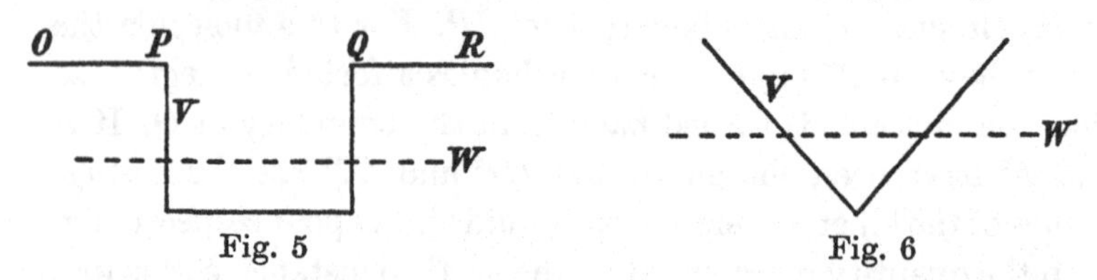

Fig. 5 Fig. 6

piece of metal. But if we alter the scale of abscissae of fig. 5, we arrive at the situation already described in fig. 2, where the width of the allowed region was only one atomic diameter. Fig. 6, which is a simplified version of fig. 2 formed from two straight lines, will be discussed in Chapter III. Here it is only necessary to add that in the same type of diagram is embraced the union of electrons and a positive nucleus to form a neutral atom. Although the electrons within an atom are in rapid motion, with kinetic energy equal to $W-V$, each is of course confined in a very small allowed region. If, for example, we plot the mutual potential energy, $V=-\epsilon^2/r$, of the electron and proton in a hydrogen atom, we have fig. 7. Any value of W lying wholly above this curve represents a free electron; for a bound electron the V- and W-curves must cut, as in fig. 7. It was an essential feature of Bohr's atomic model to suppose that for bound electrons only certain discrete values of W occurred in nature.

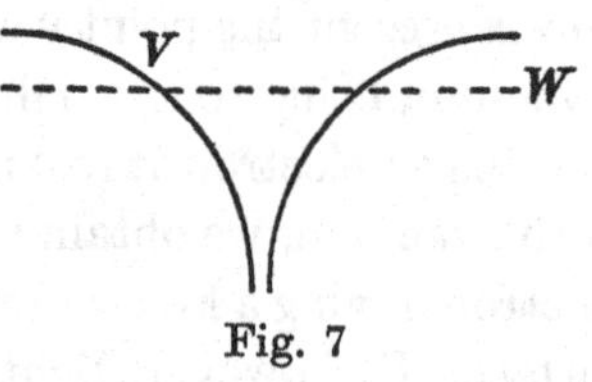

Fig. 7

§ 2. In the following chapters it will often be necessary to consider the actual magnitude of atomic and molecular energies. And this seems the most convenient point to look into the orders of magnitude involved. The spacing of the allowed energy levels in atoms and molecules determines the frequencies ν of the radiation that the system can emit and absorb, according to the relation

$$h\nu = W_m - W_n,$$

where $h = 6{\cdot}5 \times 10^{-27}$ erg secs., and W_m and W_n are any two of the energies expressed in ergs. The frequency of violet light is about 10^{15} sec.$^{-1}$; hence a quantum of violet light contains about 6×10^{-12} ergs. The most convenient unit for measuring atomic energies is the electron-volt, that is, the energy acquired by an electron or proton in falling through a potential difference of one volt; it is equal to $1{\cdot}6 \times 10^{-12}$ ergs. Hence quanta of visible light contain from 2 to 4 electron-volts of energy. (The volt is defined as 10^8 e.m.u.; dividing this by the velocity of light, namely $c = 3 \times 10^{10}$, we find that the volt is 1/300 e.s.u. The electronic charge ϵ is $4{\cdot}77 \times 10^{-10}$ e.s.u., and the value of the electron-volt in ergs, often written "ϵ-volt", is 1/300 of this.) It is useful to remember that the infra-red wave-length 12,345 Ångström units, or $1{\cdot}2345\mu$, corresponds to exactly one ϵ-volt (as may easily be verified from the relation $h\nu = hc/\lambda$). The vibrational energies of molecules are usually a little less than this, while the rotational energies give rise to spectra in the far infra-red, with wave-lengths approaching a millimetre.

It will be useful to compare these values with those of thermal energies at room temperature. The molar gas constant is $8{\cdot}3 \times 10^7$ ergs; dividing this by Avogadro's number, $6{\cdot}06 \times 10^{23}$, we obtain the atomic gas constant, $k = 1{\cdot}37 \times 10^{-16}$ ergs. According to the Boltzmann relation, in an assembly at temperature T the number of atoms or molecules having internal energy equal to E is proportional to $e^{-E/kT}$. It is useful to remember that at room temperature the value of kT is about 1/40 of an electron-volt. If then we ask how many molecules possess energy as great as one ϵ-volt, the answer is that a fraction e^{-40}, or 10^{-17}, will do so. It is only at high temperatures that an appreciable number of atoms or molecules are excited to levels as high as this. On the other hand, the spacing between the rotational levels of molecules, mentioned above, is often of the order of $0{\cdot}01$ ϵ-volt. Consequently, even at room temperature the molecules are distributed among a number of different rotational states.

CHAPTER II

It is a commonplace of scientific observation, that when a particular quantity is measured repeatedly with the highest possible precision, not a single value is obtained, but a series of values lying about a certain mean value. Values differing from the mean by a large amount are obtained less often than nearer values. If against each value obtained we plot the number of times it has been obtained, we shall have some form of the familiar "error curve", fig. 8*a*, p. 11.

When a physicist predicts the result of an atomic experiment, his prediction is often embodied in a distribution curve. His recording instruments are not usually sensitive to individual atoms, and in any case the conditions of experiment cannot be made sufficiently precise to enable him to predict a single definite value for the quantity being measured. Often the best he can do then is to express the expected result by means of a curve showing the number of particles for which the measured value will lie, say, between q and $q+dq$. To such a curve it will be convenient to give a name; we will call it the pattern of the predicted results, characteristic of the particular conditions and apparatus used.

An important innovation of quantum mechanics is that it has entirely altered our attitude to such "patterns". The former method of predicting them was as follows: (*a*) calculate what would happen under various ideally precise conditions, predicting for each ideal experiment, not a pattern, but one definite value; and (*b*) having done this, take into account the finite width of the slits in the apparatus, etc., and putting together the various predicted values in their proper proportions, obtain the required curve. The whole method was thus based on the assumption that any ideally precise measurement would lead to one unambiguous result, and that by compounding these results the required pattern was to be obtained. Quantum mechanics, however, is based on the revolutionary idea that the initial assumption is often wrong. If we imagine ourselves to be using ideal apparatus to make observations on a molecular system, we have to admit that there is often no such thing as a single *correct* value for each measured quantity; even by a series of ideally precise measurements the values obtained will have a certain residual "spread". Quantum mechanics has thus introduced the idea of purely theoretical patterns whose shape depends only on the nature of the universe and not at all on any particular apparatus or conditions of experiment.

The reasons for this innovation may be approached in the following way. To determine the position of a moving object we may direct a beam of light on to it, and look to see where it is at a given moment. The ray of light will, we know, exert a pressure on the object, but this is usually too small to disturb its motion. Unless the object has a very small mass, it will not be appreciably accelerated by the incident light. But when we come to very small particles we are faced with a dilemma. For if, on the one hand, the light is not scattered to our eye or sensitive instrument, it does not tell us where the particle was. And if, on the other hand, the radiation is scattered by the particle, there is a transfer of momentum and the particle suffers a recoil, which changes its velocity. Now it might be thought that we could make an exact allowance for this recoil, and so determine precisely both the position and velocity of the particle at a given moment. But this turns out not to be possible, even by an ideal experiment. And we know of no more delicate way of making the observation than by this method of incident radiation; all other methods are in fact clumsier. A serious oversight had thus been made in atomic theory; the fundamental fact had been overlooked, that we cannot observe an atomic system without disturbing it.

In science it is valueless to make an experiment upon a system about whose condition we have scanty information; for no useful conclusion could be drawn from it. It is essential to get the system as well as possible under control by means of preliminary observations. In the older physics it was assumed that in an ideal experiment this initial information would be obtained as a set of definite values, one for each quantity necessary to describe the state of the system at a given moment. But it is now recognised that the preliminary observations are themselves of the nature of a physical experiment, and can give only a blurred substitute for the information which a physicist used to demand; for some quantities at least, they give patterns instead of definite values. The patterns belonging to any system are not independent. As will be explained in Chapter IV, a moderately accurate knowledge of one quantity involves a proportionately greater "spread" in the pattern for some other quantity.

It was formerly assumed that small particles, such as electrons and protons, would obey the same laws of mechanics as macroscopic particles had been found to obey. But we must now admit that, for the reasons given above, atomic particles will never be found to obey the usual laws of mechanics. Since this involves abandoning to some extent the use of definite values, we have to invent a new language in which to talk about molecular systems. It might be thought that, if definite values are renounced, it would now be impossible to specify a definite state of a system. But this is not so; all we have to do is to use the patterns in place of the corresponding definite values. We have at least a definite pattern for each observable quantity. And the reader will see at once that a complete group of patterns will specify one particular state of the system. Any other state will be specified by a group containing different patterns. These patterns are the new language of physics, and it requires only a slight mental readjustment before we can use them as readily as the older language. It is the business of the theoretical physicist to find the shape of these patterns for various atomic problems, and to study how the shape of the patterns will change when he makes some alteration in the system, e.g. when he applies a magnetic field, a beam of incident radiation, or an impinging particle. When this study has been completed he knows all that he needs to know.

It will be as well to recall in some detail what are the obvious properties of any theoretical or empirical pattern. Suppose first that we have made an experiment, and that fig. 8 gives the smoothed curve embodying the results of a very large number of measurements of some quantity q, the abscissae being q. If we take a small range of values δq and construct a vertical strip, as in fig. 8, the area of this strip $N(q)\,\delta q$ gives the number of observations whose results lay in this little range δq. Since the same is true for any other strip, the total area under the curve is clearly equal to the total number of observations that have been made. If now we decide to make a few additional observations of the same quantity, say n observations, we can find at once from the curve what is the probability that any of them will lie in the range δq. We do this by altering the scale of ordinates until the total area under the curve becomes equal to n; in particular, if we were to make the area under the curve equal to unity, then the area of any strip would give us $P(q)\,\delta q$, the probability that the result of *any one* observation lies between q and $q+\delta q$. This has been mentioned here because it is the standard method of dealing with the theoretical patterns of quantum mechanics. When the area under the curve has been made equal to unity, the pattern is said to be "normalised" to unity. Of course we intend to predict the patterns from pure theory, not to obtain them from measurements.

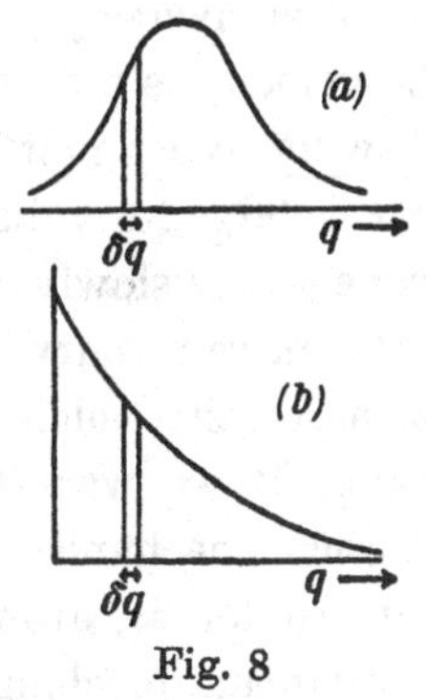

Fig. 8

The meaning of patterns can be made clearer by considering what would be the shape of some pattern according to classical mechanics. For example, we know that if at random intervals we were to take a number of instantaneous snapshots of a slowly swinging pendulum, few of the photographs would be found to show the pendulum near its central position—or rather it is unlikely that they would—simply because the swinging pendulum spends comparatively little time near its central position where its velocity is greatest. In classical mechanics the probability

$P(q)\,dq$ that the displacement of the pendulum bob will be found to have a value lying between q and $q+dq$ is easily calculated. For P is inversely proportional to the velocity at any point, and the velocity is proportional to $\sqrt{a^2-q^2}$, where a is the amplitude of swing. When plotted, P gives a curve such as fig. 9a. The value of P is large for values of q just less than a, where the pendulum travels very slowly, and P is of course zero for $|q|$ greater than $|a|$. The same curve applies to any quantity which varies simple harmonically, such as the displacement in a monochromatic wave. If we were to use patterns in classical mechanics, this would be the pattern for an atomic oscillator. But in quantum mechanics we may expect that the unavoidable "spread" will give us some blurred version of the classical pattern, as in curve b of fig. 9. We shall see in the next chapter to what extent these anticipations are fulfilled.

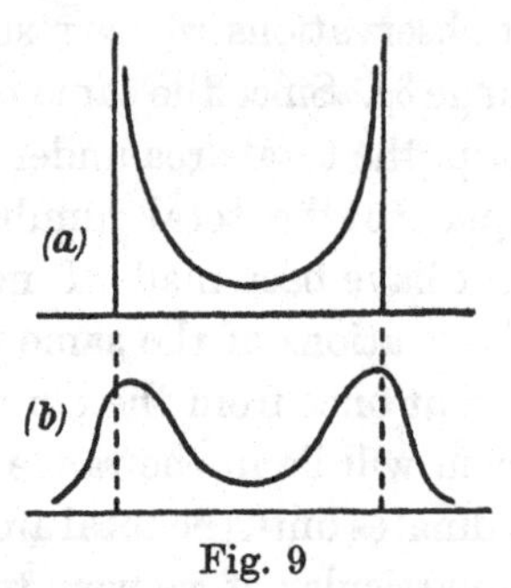

Fig. 9

Having decided to use probability patterns, what we need is a method for calculating the appropriate function for every quantity that we can measure. If we had an equation giving us $P(q)$ directly, we should have all that we require. Here, however, we come upon an unexpected feature of the theory. For the Schroedinger equation, and other equations, which have been so successful in leading to results in agreement with experiment, never give us the probability P directly, but always give us first a kind of square root of P. This is a general feature of quantum mechanics, which we could not have anticipated. It applies to every kind of pattern, of which we will give a few examples, as diverse as possible.

(1) If we are dealing with a beam of polarised light, there will be a pattern telling us the probability that the intensity of the electric field in the wave (the electric vector) will be found to have a value lying between E and $E+dE$. Instead of dealing with $P(E)$ directly, we have to say, Let $P(E)=|\psi(E)|^2$. And our equations deal with $\psi(E)$, from which we only obtain $P(E)$ by squaring the modulus.

(2) If we are considering the momentum p of a particle, and require the pattern giving the probability P that the value lies between p and $p+dp$, we have to say, Let $|\psi(p)|^2=P(p)$. And our equations give us $\psi(p)$, from which we obtain P.

(3) Suppose that we are dealing with an atom which can exist in a number of different states, which we can call 1, 2, 3, etc. Let the probability that it will be found in state 1 be denoted by P_1, the probability that it will be found in state 2 by P_2, and so on. Then we have to say, Let $|a_1|^2=P_1$, $|a_2|^2=P_2$, ..., and we have to use a_1 and a_2 in our equations instead of P_1, P_2 themselves.

(4) Suppose that we are considering two particles making a head-on collision. The pattern describing the impact will tell us the probability $P(x_1, x_2)\,dx_1\,dx_2$ that one particle will be found at a point between x_1 and x_1+dx_1 and the other particle will be found simultaneously between x_2 and x_2+dx_2. This we shall obtain in the form

$$|\psi(x_1, x_2)|^2=P(x_1, x_2).$$

Why we have to use the square roots of each probability in this way is not at all clear; the theory of quantum mechanics has been built up by guesswork, and we use this method because it enables us to obtain patterns in agreement with experiment. The use of square roots has analogies in various parts of physics: in all kinds of wave motion—sound-waves, light-waves, etc.—it is with the square root of the intensity that we always work, i.e. with the amplitude of the wave. All interference phenomena are due to the fact that intensities are only to be obtained by squaring the amplitudes. Making use of this analogy, one may speak of probability amplitudes. The equation which gives us any ψ is called a wave equation; and quantum mechanics is known as wave mechanics.

On examining the amplitudes in examples (1) to (4), it will be seen that they are of quite a different nature from the familiar waves of physics. The essential feature of sound-waves, light-waves, etc. is that they travel in ordinary space, i.e. the displacement in the wave is a function of the ordinary space co-ordinates, such as x, y, z (as well as of the time). And it will be seen that this is not true of the ψ in any of the examples mentioned; these particular amplitudes cannot then be visualised as those of waves of the familiar kind, and here the wave analogy is not helpful. Fortunately, however, there is just one class of patterns —an important class—whose ψ can be visualised in this way, since they are functions of the simple space co-ordinates. What patterns will have this property? Clearly we shall have P as a function of x, y, z—and consequently $\psi(x, y, z)$—when we are considering the probability that some single particle shall be found at or near any point x, y, z. Patterns of this kind, and the ψ-waves belonging to them, may be visualised as existing in ordinary space. To take an example—the "orbit" of an electron in a hydrogen atom will be represented by a probability pattern, and this pattern can for some purposes be visualised as a little electron cloud surrounding the positive nucleus, a little distribution of negative charge in which the proton is embedded. For each of the excited states of the atom the pattern will have a shape different from that of the ground state. Pictures of some of these patterns are reproduced in fig. 10. It is only natural that we use this simplest type of ψ-wave whenever possible. We shall not need to make use of a more complicated type until Chapter VI.

The reader will be familiar with the fact that de Broglie in 1924 suggested that waves are somehow to be associated with moving particles. To a particle of mass m, moving with uniform velocity v (that is, with kinetic energy $W - V = \frac{1}{2}mv^2$), he ascribed an associated wave-length

$$\lambda = \frac{h}{mv} = \frac{h}{\sqrt{2m(W - V)}} \qquad \text{......(1)},$$

where h is Planck's constant. This expression was derived only by analogy; like most of the ideas of quantum mechanics it was a conjecture. As will be explained in the next chapter, these wave-lengths have been incorporated into the general theory as the appropriate ψ-pattern for particles moving with uniform velocity v. For example, in the interior of the potential box of fig. 5, where $(W - V)$ is constant, the ψ-pattern will be just a simple sine curve or cosine curve with wave-length given by (1). In these patterns

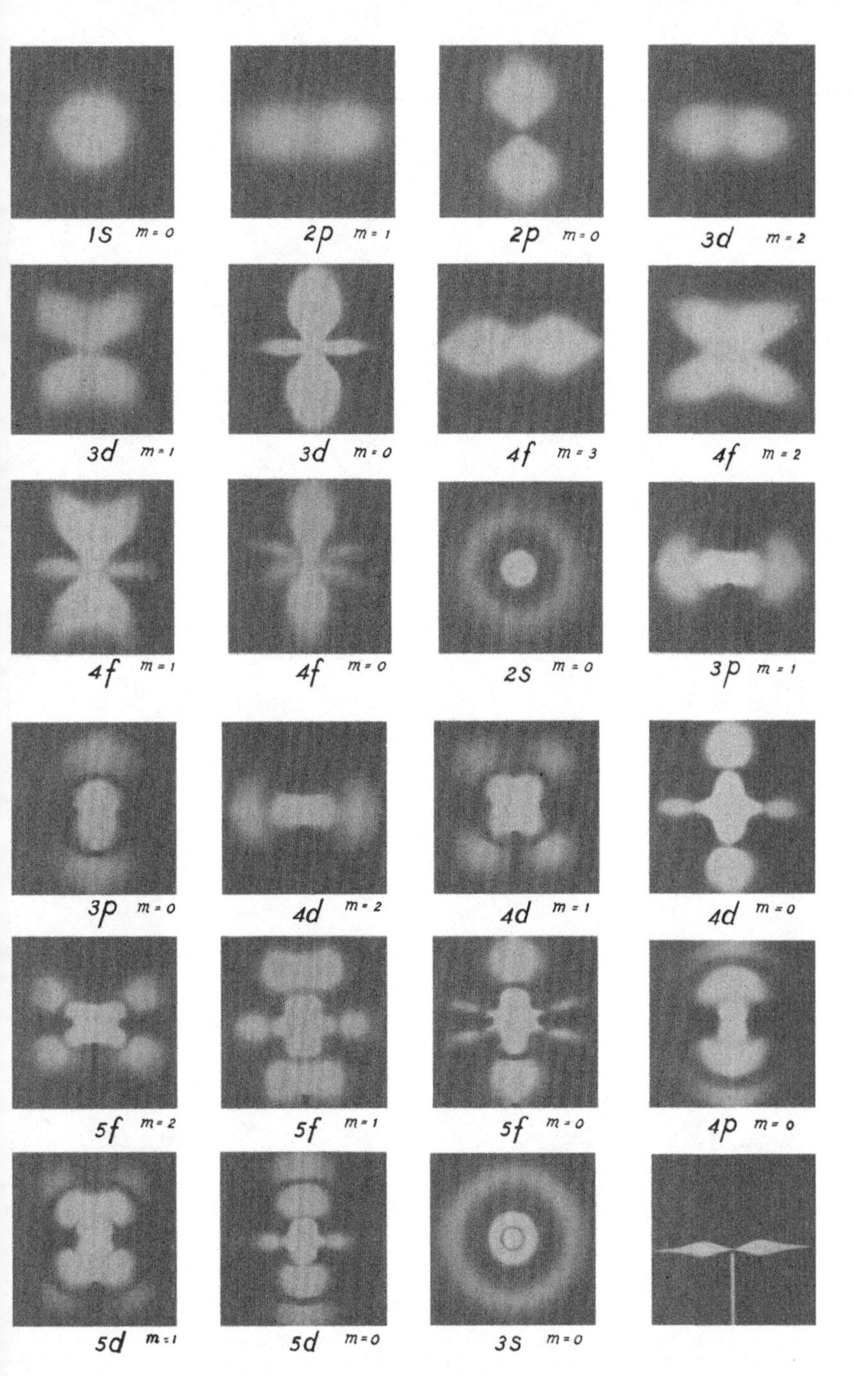

Fig. 10. Pictorial representations of the ψ-patterns or electron clouds belonging to the ground level and to some of the excited levels of the hydrogen atom

ψ of course takes negative as well as positive values; this does not lead to difficulties, since $|\psi|^2$ is necessarily positive, giving a positive value of the probability P.

According to the expression for λ, the higher the velocity of the particles the shorter is the associated wave-length. Hence where a stream of particles suffers gradual acceleration or retardation there is a continual change of wave-length. In physics we are accustomed to waves of light and sound travelling in media like air, water and glass, which are homogeneous, or nearly so. We are not so familiar with waves travelling through a medium whose refractive index varies slowly and continuously from place to place; but this is merely because it is so difficult to construct such a medium to order. We know, however, that in a medium of this kind the wave-length of any monochromatic wave would vary from point to point, as in fig. 11. With the change of wave-length is associated a change of amplitude, not shown in the diagram.

Fig. 11

Suppose that we have such a medium, and that we know how the refractive index for light of any frequency varies with x, y, z. A physicist will be able to tell by inspection the approximate nature of the wave form for any frequency. By solving the appropriate equation of wave propagation, a mathematician will be able to calculate the precise wave forms. It is just this type of problem and this procedure which most nearly resemble the methods, qualitative and exact, of tackling an atomic problem in wave mechanics. For in any atomic problem we are given certain forces of attraction and repulsion between the particles, with which there is associated a certain potential energy V. When we are given the V as a function of x, y, z, as in the problems of Chapter I, this is equivalent to being given the value of the refractive index of the medium at all points. By solving the Schroedinger wave equation, the mathematician can sometimes calculate the precise value of ψ at all points. But for a physicist or chemist it is also important to be able to tell by inspection the rough form of the ψ, and hence approximately how the system will behave.

CHAPTER III

§1. THE WAVE EQUATION

Having discussed the nature of probability patterns for atomic systems, we may attack the problems of Chapter I. We saw there how in classical mechanics the potential energy V was responsible for keeping electrons and atoms in their places. Each particle in the universe is moving in a region where $V < W$; in whatever direction it is moving, it comes sooner or later to a surface where $V = W$, and here it is turned back. We shall not expect to find in quantum mechanics anything so definite as this sharp dividing surface. And it is of the first importance to decide what shall replace it.

Schroedinger hit on the idea that an equation whose solutions are de Broglie waves where $W > V$ might also give reasonable results where $V > W$. Such an equation is

$$\frac{d^2\psi}{dx^2} + \frac{8\pi^2 m}{h^2}(W - V)\psi = 0 \qquad \text{......(2).}$$

For when $(W - V)$ is positive and either constant or varying slowly with x, this equation has as solution

$$\psi(x) = A\cos 2\pi\frac{x}{\lambda} + B\sin 2\pi\frac{x}{\lambda} \qquad \text{......(3),}$$

where A and B are constants, and λ is the de Broglie wave-length given by (1), as may easily be verified by differentiating (3) with respect to x. Thus so long as $W > V$ and V varies slowly, equation (2) is merely the mathematical expression of the fact that ψ has a wave-length $h/\sqrt{2m(W-V)}$. Schroedinger went on to suppose that the equation was still applicable for $W < V$, and for V varying quickly with x. Solutions of (2) in fact turn out to be patterns of a type which we require, namely those whose $|\psi|^2$ gives the probability $P(x)$ when W is known.

Let us apply equation (2) to the problem of a particle in the potential box of fig. 5 on p. 6, in which the lines PO and QR are supposed to extend to infinity in either direction. The solution (3) is valid in the interior of the potential box for any W greater than V. In the forbidden region on either side we have $(W-V)$ constant and of negative sign. The solution of (2) now is

$$\psi(x)=Ce^{-kx}+De^{kx} \qquad \text{......(4)},$$

where C and D are constants, and

$$k=\frac{2\pi}{h}\sqrt{2m(V-W)} \qquad \text{......(5)},$$

as may be easily verified by differentiating (4).

We see that ψ contains a term which decreases exponentially and a term which increases exponentially. Evidently the coefficient of the latter must be zero if the pattern is to represent a particle bound in the potential box; otherwise ψ and ψ^2 would become infinite, representing a particle which spent its time at an infinite distance from the laboratory. To the right-hand side of the potential box D must be zero, and to the left, where x goes to $-\infty$, C must be zero. On each side then we are left with the decreasing exponential only; and if k is sufficiently large we have a ψ whose value falls off rapidly as we penetrate into the region beyond the classical boundary where $W=V$. The pattern for fig. 5 is to be obtained by joining the three portions together—in the middle is the sine curve or a portion of one, and exponential tails stretching out indefinitely in either direction. Curve a of fig. 12 gives a possible shape of the pattern for a particular value of W;

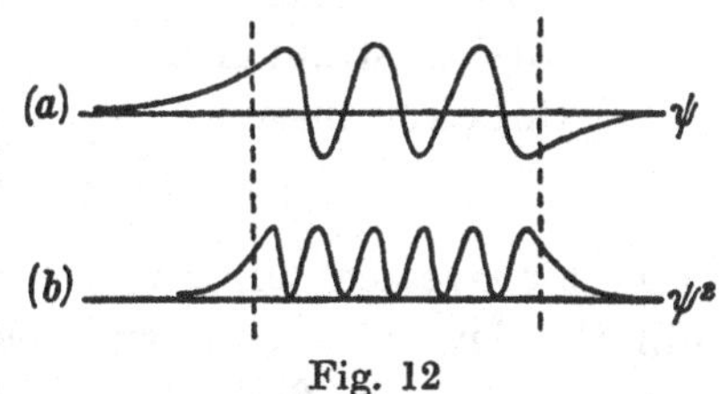

Fig. 12

and curve b gives the form of the corresponding ψ^2. When the ordinates of curve a have been chosen so that the total area under curve b is equal to unity, the pattern is said to be normalised.

§2. Until the 'eighties of the last century the nature of the electric current which flows through metals was unknown. It was provisionally supposed that positive electricity was flowing in the direction called positive, while perhaps negative electricity was simultaneously flowing in the reverse direction. Finally it was discovered that only one kind of electricity was in motion, namely electrons streaming in the direction which had unluckily been called negative. When a current passes through a series of metals in contact, the cores of the atoms remain in position, while the electrons stream freely from one metal to the next in the circuit. If then it comes to modifying the classical laws of dynamics, we shall perhaps be prepared to admit a greater measure of freedom to electrons than to atoms.

Now we have seen in the preceding section that, if we use patterns derived from equation (2), the classical division into allowed and forbidden regions disappears, for $|\psi(x)|^2 dx$ is the probability of finding the particle in any range dx. A particle approaching and penetrating beyond the former boundary will be turned back sooner or later, but there is no limit to the distance it *may* go before returning. Instead of a complete ban beyond the classical boundary we have ψ^2 decreasing at the rate e^{-2kx}, which we must now evaluate. We need the following quantities:

Planck's constant:	$h = 6{\cdot}5 \times 10^{-27}$ ergs secs.
Mass of electron:	$m = 9{\cdot}0 \times 10^{-28}$ grams.
Mass of proton:	$M = 1{\cdot}6 \times 10^{-24}$ grams.
One Ångström unit	$= 10^{-8}$ cm.
One electron-volt	$= 1{\cdot}6 \times 10^{-12}$ ergs.

$$\log_{10} e = 0{\cdot}4343.$$

The most convenient units to use for atomic systems are for length the Ångström unit, and for energy the electron-volt. Using these units we find

$$2kx = 3 \times 10^{13} \times \sqrt{m} \, . \, \sqrt{V - W} \qquad \text{......(6).}$$

Choosing any typical atomic energies we can use this expression to find how efficient the factor is in providing a substitute for the classical boundary. Let us for example set $(V - W)$ equal to one electron-volt, and calculate the value of e^{-2kx} at a distance of

2 Ångström units beyond the classical boundary. We find approximately:

for electrons $\quad e^{-2kx} = e^{-2\cdot 3} = 0\cdot 1,$

and for protons $\quad e^{-2kx} = e^{-100} = 10^{-43}.$

We see that there is an acute discrimination between electrons and protons. For electrons the value of ψ^2 does not decrease rapidly, and their behaviour will be quite different from that predicted from classical mechanics. For protons, on the other hand, there is a fairly efficient substitute for the classical boundary, which will be still more efficient for heavier particles; it was for this reason that the inadequacy of classical mechanics was not discovered earlier. It will be noticed that the electric charge borne by the particle does not occur explicitly in equation (2). It enters, however, through the potential energy V, since in any electric field V will depend on the value and sign of the charge carried.

§ 3. We may now obtain the appropriate pattern for the problem of fig. 1 and fig. 3 on pp. 1 and 4, where we had particles retarded and brought to rest by a constant force. The potential energy is linear, hence for $(W-V)$ in equation (2) we must write $(W-cx)$, where c is a constant. The solution of the equation involves a Bessel function. Of the two terms we again reject the one which becomes infinite, using the other solution which tends to zero at an infinite distance. The resulting pattern for ψ and for ψ^2 in the interesting region near the classical boundary is plotted in fig. 13 for a particular slope of the V-curve. The steeper the slope, the more rapidly will λ change. Where the particle is being retarded the amplitude increases (as we should expect from the argument given on p. 12), and beyond the classical boundary, represented in fig. 13

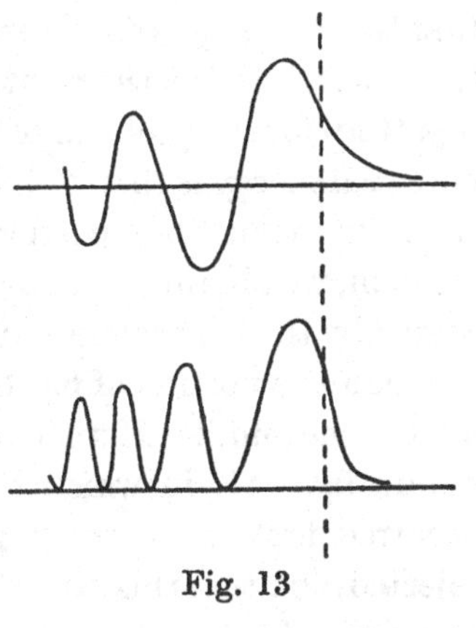

Fig. 13

by the vertical dotted line, the value of ψ falls off exponentially, as in the previous problem. There is again a small probability of the particle being found a long way beyond the classical turning point. The exponential rate at which ψ falls off again depends violently on the mass of the particle; and we may say at once that a calculation similar to that given above applies to any form of V-curve which provides a boundary.

We find then that we can safely abandon classical mechanics and use equation (2) instead. Although the division into allowed and forbidden regions seemed to be essential to preserve the stability of matter, we obtain a qualitative substitute which is just sufficiently rigorous to control heavy atoms, while allowing considerable freedom to electrons, and in a smaller degree to protons. The programme of quantum mechanics involves then nothing less than the re-examination of the whole of atomic and molecular physics, using the new mechanics to describe the behaviour of particles.

§ 4. We may notice that in the problem of § 3 the ψ-pattern has the same form for all values of the energy W. For if in fig. 1 we draw a horizontal line to represent some other value of W, we see that the values of $(W-V)$ are the same as before, only all shifted to the left, or all to the right. We may use this fact to obtain the ψ-pattern for the problem of fig. 6 (p. 6), in each half of which the potential energy is that with which we have just dealt. In each half the ψ-curve must be a portion of fig. 13. Choosing a value of W at random, we obtain a boundary on either side; let these be represented by the vertical dotted lines in fig. 14. Fitting back to back the requisite portion of fig. 13 we obtain fig. 14*a*, with an absurd kink in the middle. If we invert one half, which is permissible, we obtain fig. 14*b*, in which the curve again fails to join up. It occurred first to Schroedinger that such solutions were to be rejected, and that this leads directly to quantisation of the energy. For if with fig. 6 we had chosen a slightly lower value for W, with consequently a narrower allowed region, the maxima which fail to coincide in the centre of fig. 14*a* would have fallen slightly nearer together; and we can clearly find a particular value of W

for which the two halves of the curve join up to give a satisfactory pattern. This, Schroedinger suggested, will be one of the allowed quantised levels. In the same way we can find a certain higher energy for which the allowed region will have exactly the right width to enable the two halves of fig. 14*b* to join up. By sliding the two halves in fig. 14 relative to one another in this way, we obviously obtain an unlimited number of satisfactory ψ-patterns corresponding to discrete energies spaced at intervals. The use of equation (2) to obtain probability patterns thus yields as a by-product a set of discrete quantised levels, a welcome result.

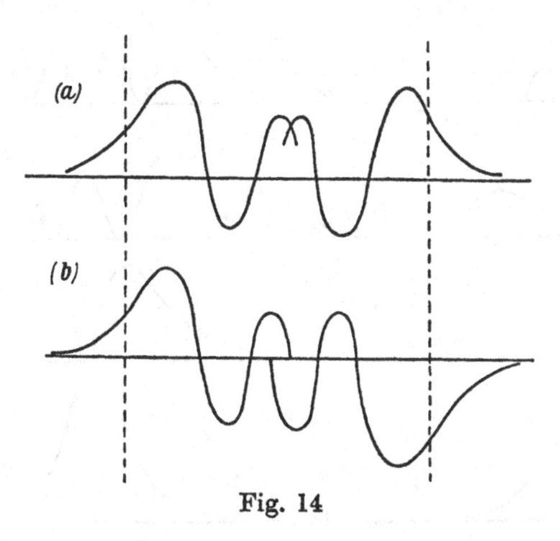

Fig. 14

This conclusion is fundamental, for it applies to any kind of potential box, whatever the form of the V-curve. In offering fig. 12 as a possible solution for a rectangular potential box, we refrained from pointing out the difficulty of fitting the exponentials on the de Broglie sine curve. In general this cannot be done without a kink somewhere; for, the length of the box being given in advance, there are only certain wave-lengths which just fit into it and join on smoothly to the required exponentials at both ends of the box. ψ and $d\psi/dx$ must be made continuous everywhere. (This could always be done if we were allowed to use the mixture of increasing and decreasing exponentials of (4), but the value of ψ would become infinite and the result would be meaningless.) As in the previous problem there are only certain discrete energies, spaced at intervals, which have acceptable patterns. The lowest allowed energy is one possessing a de Broglie wave-length more than twice the length of the potential box, so that the pattern consists of the two exponentials with less than half a wave of the sine curve fitted in between, fig. 15*a*. Clearly it is impossible to obtain a continuous ψ-curve for any energy lower than this. The next higher level belongs to an energy for which nearly a whole wave-length fits into the potential box, fig. 15*b*, and so on. Returning to fig. 14, we see that the lowest patterns there will be similar, although

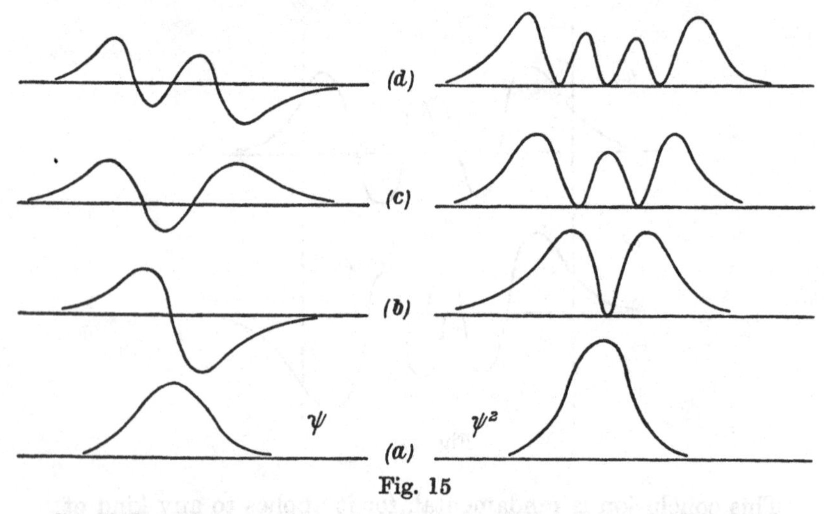

Fig. 15

they belong to a potential box of entirely different form. The lowest level is obtained from small portions of fig. 14*a* and has a pattern like fig. 15*a*; the next level is obtained from fig. 14*b* with a pattern like fig. 15*b*; the next from fig. 14*a* with a pattern like fig. 15*c*, and so on.

For a reason which will be explained below, the main characters of the set of ψ-patterns is quite independent of the form of the potential box. The ψ-curves always resemble the normal modes of vibration of a piano wire. The ψ-curve for the lowest level is always without any node, fig. 15*a*; the next always has one node, as in fig. 15*b*; the third with two nodes, and the *n*th with $n-1$ nodes. The curves plotted in fig. 15 are, as a matter of fact, the solutions for a potential box whose V-curve is a parabola, as in fig. 2, that is, for a harmonic oscillator. The curve for the sixth allowed level will resemble fig. 12, except that the amplitude of the sine curve will not be uniform across the box, and the five zeros will not be quite equally spaced. It is only to this extent that the patterns for any type of potential box differ from one another.

It is easy to see why the ψ-curves appropriate to different potential boxes resemble one another so closely. For $d^2\psi/dx^2$ is the rate of change of slope of the curve, i.e. the curvature; and equation (2) may be written

$$\frac{d^2\psi/dx^2}{\psi} = -C(W-V) \qquad \text{......(7)}.$$

Where $W > V$, both sides of this equation are negative, and the curve must everywhere keep its concave side towards the x-axis, as it always does, for example, in sine and cosine curves, etc. Where $W < V$, on the other hand, the curve must keep its convex side towards the x-axis, as in exponential curves. The classical boundary, where $W = V$, must always mark the division where the character of the ψ-curve changes over from one to the other. Thus in solving any problem we always have in advance a good idea of what the forms of the possible patterns are going to be as soon as we have drawn an energy diagram.

In the following chapters, when confronted by the problem of finding the allowed energies of a system, we shall refer back to the programme carried out here. The method consists of "fitting" ψ-curves into the potential box. The pattern for the lowest level is always without any node; this is the most important level, because the system cannot have an energy lower than this. The next higher allowed level will be that energy whose pattern will fit into the box with one node. In looking for successive higher levels, we are using successively shorter wave-lengths, and (except for the artificial V-curve of fig. 5) at the same time the potential box is getting wider. The spacing of the levels, i.e. the difference in energy between one level and the next, depends upon the size and shape of the potential box. Clearly it may happen that for a particular kind of V-curve the successive energy levels are exactly equally spaced. This in fact happens for the harmonic oscillator, in which the potential energy is proportional to the square of the displacement from the mean position, $V=\frac{1}{2}\alpha x^2$. The allowed energies to which the series of patterns of fig. 13 belong are given by

$$W_n=(n+\tfrac{1}{2})\frac{h}{2\pi}\sqrt{\alpha/m} \qquad \text{......(8),}$$

where n takes the integral values 0, 1, 2, ... for successive levels and is known as the quantum number. The set of levels thus has uniform spacing throughout, the interval being $\frac{h}{2\pi}\sqrt{\alpha/m}$. In fig. 7, on the other hand, the width of the potential box increases at first slowly and then rapidly, and becomes infinite; consequently the levels are at first widely spaced, the higher levels are more and more closely spaced, and finally converge to a limit.

At the end of Chapter I attention was drawn to the fact that the energy levels in an atom are spaced about a hundred times as widely as those in molecules. Patterns like those of fig. 15 will be used to represent an electron in an atom or an atom in a molecule, but the wave-length will be very different, since it depends on the mass of the particle by the relation $\lambda=h/mv$. What then will be the spacing in each case between the levels to which these patterns will belong? It will be convenient to show here numerically that the spacing is going to be of the right order of magnitude. For electronic levels, de Broglie wave-lengths differing by a factor of 2 or 3 should correspond to energies differing by several ϵ-volts. For an electron in an atomic box of width 3 Ångström units we will take $\frac{1}{2}\lambda=3\times10^{-8}$ cm., and estimate the kinetic energy from (1):

$$(W-V)=\frac{h}{2m\lambda^2}=5\times10^{-12}\,\text{ergs}=3\,\epsilon\text{-volts}.$$

For a proton or heavier particle, on the other hand, the spacing of the levels will be closer, even when the width of the potential box is considerably less.

§5. THE HYDROGEN ATOM

Steps must now be taken to extend these methods to problems in three dimensions. Consider once more the problem of fig. 1. Our particle in fig. 1 rose to a height corresponding to the point Q, the height at which $W-V=0$; this was because we had stipulated that the particle had been projected vertically. A particle projected at an angle to the vertical, with the same energy W, would not rise to so great a height, because part of the kinetic energy is associated with the horizontal component of the motion. To find how high the particle will rise, one has to subtract this part of the energy, and in this way one obtains the classical boundary. The procedure of finding the appropriate ψ-pattern is the same as before. With rectangular axes, x, y, z, one has to use in place of equation (2)

$$\frac{d^2\psi}{dx^2}+\frac{d^2\psi}{dy^2}+\frac{d^2\psi}{dz^2}+\frac{8\pi^2 m}{h^2}(W-V)\psi=0 \qquad \text{......(9)},$$

where the kinetic energy

$$W-V=\frac{m}{2}(\dot{x}^2+\dot{y}^2+\dot{z}^2) \qquad \text{......(10)}.$$

A three-dimensional potential box will be a region where $W>V$, surrounded on all sides by a forbidden region where $W<V$. To this an argument will apply similar to that given with reference to equation (7). Inside a certain surface the ψ-pattern will have one character, and outside the surface the value of ψ will die away exponentially in all directions.

Since a charged particle, such as an atomic nucleus, has a spherically symmetrical field, we are most interested in spherical potential boxes. Corresponding to figs. 5, 6 and 7 there will be three-dimensional problems in which the V-curves of figs. 5, 6 and 7 give the potential energy along any diameter. We shall not discuss the problems of figs. 5 and 6 further, but fig. 7 becomes the very important problem of the hydrogen atom. The proton provides a potential box for the electron; and for the classical electron orbit one must substitute the idea of a pattern. For some purposes this pattern may be thought of as an electron cloud, whose density is proportional to $|\psi|^2$, and thins out rapidly in all directions. This density is a "probability density", since in the hydrogen atom the whole ψ-pattern must be normalised to give exactly one electron, i.e. the value of ψ must be adjusted so that the integral over all space is equal to unity:

$$\int \psi\psi^* dv = 1 \qquad \text{......(11).}$$

The electron occasionally makes long excursions from the nucleus, but the probability of finding the electron far outside what we ordinarily regard as an atomic volume is very small. If we multiply ψ^2 at every point by ϵ, the electronic charge, we may picture the cloud as a distribution of negative electricity, the total quantity when integrated over all space being ϵ. When normalised as in (11) ψ^2 must have a definite numerical value at every point, and we may consider in what units it may be expressed. Its dimensions must clearly be the reciprocal of a volume, in order that (11) shall be a pure number. The order of magnitude of the normalised ψ for an atomic electron may be obtained by a rough calculation. If the atomic volume inside which ψ is appreciable is, say, 10^{-22} cm.^{3}, we could obtain unity by multiplying this by 10^{22} cm.$^{-3}$; this value would therefore be a suitable mean value for ψ^2 in the atom; and its square root, 10^{11} cm.$^{-3/2}$, would be a satisfactory mean value for ψ. As a matter of fact, in the pattern belonging to the lowest level of the hydrogen atom it is found that ψ has the value 5×10^{11} cm.$^{-3/2}$, at the centre, and falls away exponentially from this value in all directions, as in fig. 21*a*, p. 36. (One would not expect ψ in this case to have the form of fig. 15*a*, since the positive nucleus is of course situated in the centre of the potential box.) This simplest type of electron cloud, of which a pictorial representation is given in fig. 10, is very different from a classical orbit in which the motion of the electron is confined to a plane. In most cases the ψ-pattern belonging to an excited state shows a greater resemblance to an orbit, and before going further it will be convenient to look into the part played by angular momentum in the latter.

Consider, for example, a rotating elliptical orbit, according to the older ideas. The electron never goes nearer to the nucleus than *OP*, fig. 16, nor farther from it than *OQ*. Values of r between

$r = OP$ and $r = OQ$ are the allowed values of r, while $r > OQ$ and $r < OP$ define forbidden regions; which implies an effective V-curve of a form such as $ABCD$ in place of the Coulomb curve EFG, which is a plot of $-\epsilon^2/r$, as in fig. 7. Returning to the orbit PQ, we see that at Q the radial velocity is zero, but the particle still possesses an amount of kinetic energy represented by AF. The boundary at D is also to be expected, for it is of course the angular momentum about O which prevents the electron from colliding with the nucleus; and in determining the allowed values of r the kinetic energy associated with this angular momentum must be subtracted from the total kinetic energy, $W - V$. Taking any vertical line HBG, HG represents the total kinetic energy; let BG be the angular part, and HB the radial part. As H approaches A or D, HB tends to zero, giving a boundary at A or D. In this way we obtain a potential box $ABCD$ in place of the original potential box. We shall see later how this idea enters into the quantum mechanical treatment.

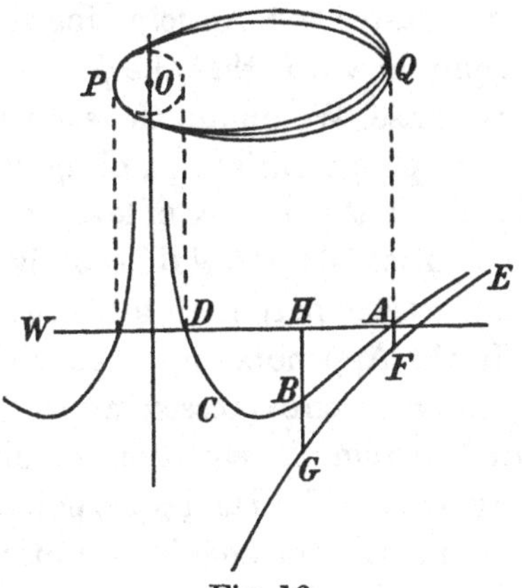

Fig. 16

§ 6. The alkali atoms Li, Na, K, Rb, Cs are said to be hydrogen-like; each consists of a positive core and a single valence electron. The core as a whole bears a single positive charge, since it contains $Z-1$ electrons surrounding a positive charge $Z\epsilon$. The potential box for the valence electron is provided by the core instead of by the nucleus. The valence electron spends most of its time outside the core, and consequently in a field almost identical with the field of a proton. For reasons which will appear later, it is convenient to carry on the discussion of these atoms at the same time as that of the hydrogen atom itself. To account for the complicated series spectra emitted by the hydrogen-like atoms, Bohr postulated the existence of several series of energy levels, of which the most important are known as the *s*-series, the *p*-series, and the *d*-series.* Anticipating the results, we may at once state roughly how it is that the ψ-patterns belonging to these various series arise. We obtain the *s*-series by "fitting" ψ-patterns into the simple potential box of fig. 7, or the curve *EFG* of fig. 16. We obtain the *p*-series of levels by fitting a set of ψ-patterns into a potential box like *ABCD* of fig. 16; and the *d*-series by fitting another set of patterns into another potential box of the same type.

In the Appendix on p. 148 it is shown that the Schroedinger equation, when expressed in spherical polar co-ordinates, splits up into three simpler equations involving r, θ, and ϕ, separately. In the equations for θ and ϕ the potential energy does not occur, and consequently the angular variation of ψ which we shall find will apply to all fields of spherical symmetry. Thus, although the discussion to be given below is directed primarily to the hydrogen atom only, it is worth bearing in mind that it has wider applications.

The simplest of the three equations is (84), of which the solution is

$$\Phi = A\cos m\phi + B\sin m\phi \qquad \text{......(12).}$$

In the special case of $m=0$ we see that Φ is constant and there is no variation with ϕ. Since the value of ϕ must lie between 0 and 2π, the value of the normalising factor is evidently $A = 1/\sqrt{2\pi}$. When m is not zero, the value of ψ by (12) varies sinusoidally with the angle ϕ, and this has important consequences. For in fig. 17*a* let *ABC* be a circle in the equatorial plane with the atomic nucleus as centre. At any point *A*, chosen at random, ψ will have a certain value. Starting from *A*, let us go round the circle watching the sinusoidal variation of ψ. When we come back to the point *A*, ψ must obviously come back to the value

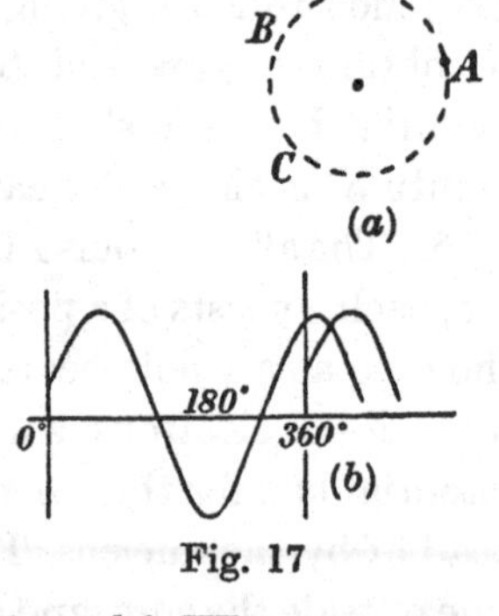

Fig. 17

* The letters *s*, *p* and *d* are the initial letters of "sharp", "principal" and "diffuse"—terms formerly used for classifying spectrum lines, and later transferred to the levels which were believed to give rise to these series of spectrum lines.

from which we started, since we are merely looking at the same thing twice. This imposes a restriction on the possible forms of ψ. For example, fig. 17*b* shows an impossible form. The portion of the curve between 360° and 450° must of course be identical with the portion between 0° and 90°, since it is recording the same thing twice. But in fig. 17*b*, with a value of m chosen at random, the curve fails to join up, and is to be rejected; the situation is similar to that of fig. 14. We can, however, find certain discrete solutions which are satisfactory, namely when m is an integer; for only then is it everywhere true that $\cos m(\phi + 2\pi) = \cos m\phi$.

A similar argument applies to the curves giving the variation of ψ with θ—see fig. 23. As mentioned in the Appendix, satisfactory patterns are obtained only when l takes the values 0, 1, 2,...; and there is the further restriction that m must not be greater than l nor less than $-l$. When $l = 0$ there is no variation with θ, and this combined with the constancy of Φ gives patterns with complete spherical symmetry. For other values of l one has various curves analogous to fig. 17.

This restriction to integral values of l introduces an angular quantisation similar to that which had previously been assumed by Bohr. He had supposed that only those orbits occurred in nature whose angular momentum was equal to $lh/2\pi$, where h is Planck's constant, and l is any integer. Writing μ for the mass of the electron, the amount of kinetic energy associated with this angular momentum is of course $\left(\frac{lh}{2\pi}\right)^2 \Big/ 2\mu r^2 = l^2h^2/8\pi^2\mu r^2$. This brings us back to the question of the potential box $ABCD$ of fig. 16, which we expected to obtain by subtraction from the total kinetic energy. If Bohr's assumption still held, we should expect that in the radial equation the kinetic energy $(W - V)$ would be reduced to

$$(W - V - l^2h^2/8\pi^2\mu r^2).$$

This is, in fact, the form which the Schroedinger equation (87) is found to take, with the exception that $l(l+1)$ occurs in place of l^2. This at once strongly suggests that Bohr's original idea of angular quantisation was very nearly correct, and that the l

which was introduced into equation (86) will play the usual rôle of an angular quantum number, while m will measure the time-average of the projection of this angular momentum on the axis. This view is supported by the forms of the patterns. For, recalling the case of linear momentum, in which we had the relation $\lambda = h/\mu v$, we should expect a similar relation for angular momentum. And we see, in fact, that the larger the value of m or l, the shorter is the wave-length in figs. 17 and 22. When in Chapter v we have considered how a flow of current is to be described in quantum mechanics, it will be possible to confirm these ideas by showing that to any electronic state there is to be ascribed a continuous flow similar to the circulation of an electron in an orbit, and proportional to l.

§ 7. Before looking into the details of any particular pattern, we will first consider how the various series of energy levels arise. Write

$$V - l(l+1)\, h^2/8\pi^2\mu r^2 = V_l \qquad \text{......(13).}$$

Then equation (87) takes the form

$$\frac{d^2F}{dr^2}\Big/ F = C\,(V_l - W) \qquad \text{......(14).}$$

The character of the F-curve depends upon the sign of $(V_l - W)$, as in equation (7). Any region where $V_l < W$ denotes a potential box into which F-curves may be fitted. When $l = 0$ the potential energy is the original potential energy of the electron in the field of the positive core or nucleus. It will be noticed that the question of the exact form of this field now arises for the first time. All the results that have been obtained so far are independent of $V(r)$, since $V(r)$ does not occur in equations (82) and (83). In the atomic problem $V(r)$ approximates to, or is equal to $-\epsilon^2/r$.

To obtain a curve for (13) we may plot the two terms separately and add the ordinates. When r is one Ångström unit, the value of $h^2/8\pi^2\mu r^2$ is 6×10^{-12} ergs, while the value of ϵ^2/r is $2{\cdot}3 \times 10^{-11}$ ergs. In fig. 18 the lowest curve is a plot of the Coulomb term, while the two dotted curves are plots of the other term for $l = 1$ and for $l = 2$. Adding the ordinates of each of these curves to the Coulomb energy we obtain the curves labelled P and D, which are similar to the curve *ABCD* of fig. 16, p. 27. For $l=3$, etc., we shall obtain less important curves lying just above the curve *D*. For large values of r all the curves become symptotic to the axis. Any horizontal line above the axis cuts a curve such as *P* once only, and represents the energy *W* of a free electron. Any horizontal line below the axis cuts a curve such as *P* twice, if at all, and thus represents the energy *W* of an electron confined in the potential box. The allowed region in space is of course obtained by rotating about *O* as centre; the region where $W > V$ is thus a spherical shell whose width depends on *W*.

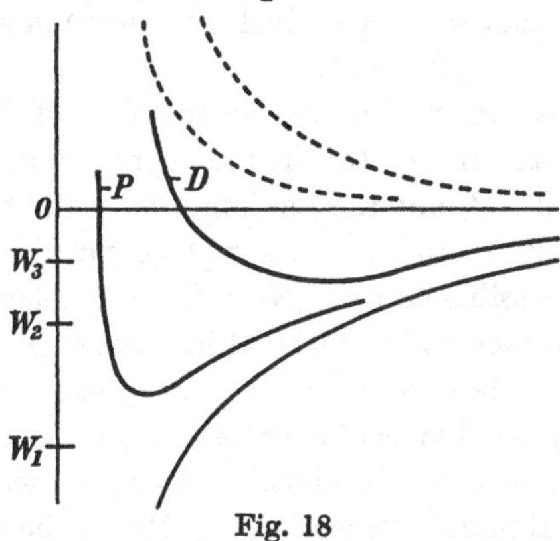

Fig. 18

By fitting solutions of (14) into the potential box labelled *P*, we obtain in the usual way a set of levels—in fact, the whole series known as the *p*-levels. This set of patterns will have the same features as those of § 4. That belonging to the lowest *p*-level will resemble curve *a* of fig. 15, with its maximum lying inside the box *P* of fig. 18. For the next higher *p*-level the value of ψ will (as in curve *b* of fig. 15) for a certain value of r inside the box be equal to zero. Since this is true at all angles, it means that ψ is zero over the whole surface of a sphere of a particular radius. For the next higher *p*-level (as in curve *c* of fig. 15) the ψ will be zero for two values of r inside the box, thus giving two spherical nodal surfaces, and so on. The same remarks apply to the potential box labelled *D*; the lowest *d*-level will have no spherical nodal surface, and so on.

For $l=0$ we are left with the unmodified Coulomb spherical box of fig. 7. In contrast to fig. 16, the nucleus is not in this case situated in a forbidden region; there is nothing to prevent the electron from frequently colliding with the nucleus. Yet the Schroedinger equation gives a set of acceptable patterns, namely those of the important *s*-series, which includes, as we shall see, the ordinary ground level of the hydrogen atom. For these levels, with $l=0$, there is no angular momentum; the electron moves along any diameter, passing right through the nucleus. It will be seen that "*s*", "*p*" and "*d*" are merely names given to the levels with $l=0$, $l=1$, and $l=2$, respectively. We have in addition less important series for $l=3$, 4, etc.

To facilitate discussion, three specimen values of the energy W have been indicated in fig. 18. In the energy range near W_2 a horizontal line will cut only the Coulomb curve and the *P*-curve; in this range only *s*- and *p*-levels will be possible. Near W_3 *d*-levels will be possible as well. Near W_1 only *s*-levels are possible; there is, in fact, only one level in this range—the lowest level of the atom, whose pattern has already been referred to in § 5. The normal ground level of every hydrogen-like atom is thus an *s*-level. The first excited level will be either the second *s*-level, with one spherical nodal surface (see fig. 10), or the first *p*-level, whichever of these happens to be the lower; in the alkali atoms the first *p*-level is lower than the second *s*-level. The next higher level than these will be the second *p*-level, the third *s*-level or the first *d*-level, and so on. All the higher levels are crowded into the range of energies between these levels and the limiting energy at which the electron becomes free (the horizontal axis in fig. 18). This limiting energy provides the usual zero from which to measure the negative energies of the atomic levels. The negative energy of the lowest quantised state, measured from this zero, is of course the ionisation potential of the atom.

In the hydrogen-like atoms the positive core provides for the valence electron a potential box which differs from $-\epsilon^2/r$ in a way characteristic of the element. Hence for each element we have the same sets of levels occurring at energies peculiar to the particular atom; the ionisation potential and the energies of the excited levels have characteristic values. The qualitative discussion of fig. 18, however, applies to all, and we can see how a system of levels arises like that shown in fig. 19, for example, which is part of the empirical scheme of levels which enables us to account for every line in the optical spectrum of sodium. A scale of volts is given, and every spot represents an energy level. The lowest level is an s-level, and we see how at higher energies the p-, d- and f-levels are added in turn, the order being p, then d, and then f, as we expect from fig. 19. (The many upper levels have been omitted from fig. 19.)

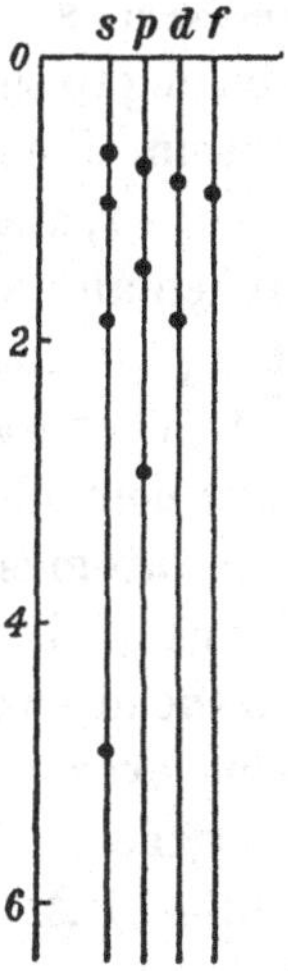

Fig. 19

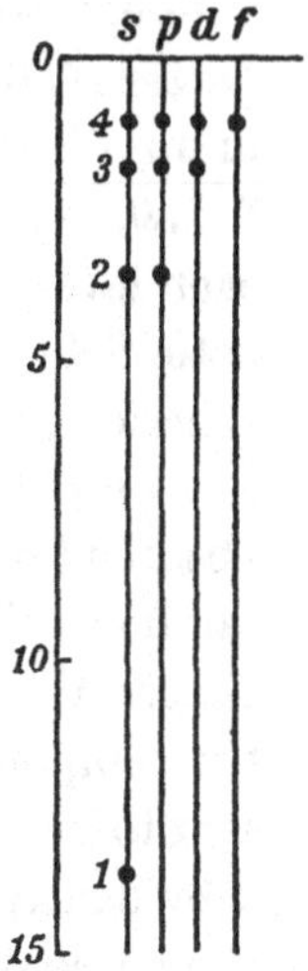

Fig. 20

In the problems of § 4 there was for each allowed energy just one possible ψ-pattern. But we see here that it *might* happen that for a particular type of the nuclear field we should find the first p-level coinciding with the second s-level; we should then have more than one ψ-pattern for a single allowed energy. This, in fact, does happen for the hydrogen atom itself. Not only that, but the energies of the third s-level, the second p-level, and the first d-level all coincide, and so on through the scheme of levels, as in fig. 20. This systematic coincidence is a property of the exact inverse-square field; the allowed levels given by the Schroedinger equation coincide, in agreement with the empirical scheme of levels, fig. 20.

In the case of the valence electron of the alkali atoms the departure of the atomic field from the inverse-square law has the effect, shown in fig. 19, of completely destroying this systematic coincidence of levels possessing different values of l. It does not however spoil the coincidence of levels with different values of m. As mentioned above, the restrictions on the values of the quantum number m are that it shall be integral, and neither greater than

l nor less than $-l$. For p-levels, where $l=1$, m can have the three values 1, 0, and -1, and to each of these values belongs a different ψ-pattern as shown for hydrogen in fig. 10. These three patterns all belong to the same value of the allowed energy, even in an alkali atom. For d-levels, where $l=2$, m takes the values 2, 1, 0, -1, and -2, and there are thus five different patterns for each level. When an atom is placed in a strong electric or magnetic field, the potential box for the electron is no longer quite spherical, and the energy levels belonging to these various patterns no longer coincide. The frequency of the light emitted or absorbed by the atom is no longer the same, and hence arises the familiar splitting of the spectrum lines in the Stark and Zeeman effects. When the intensity of the applied field is reduced to zero, the component levels close up again into a single level, which is said to be "degenerate". For any value of l the "multiplicity" is evidently $(2l+1)$. All s-levels are single,* for m can only have the value 0.

The coincident energy levels of hydrogen are numbered up from the bottom, 1, 2, 3, ..., as shown in fig. 20. These numbers are known as the principal quantum numbers, and define the energy of the levels by expression (19) in a way analogous to (8). In this notation the lowest p-level, whose ψ-pattern has no spherical nodal surface, is not called $1p$, but it is called $2p$ because it belongs to the level with principal quantum number $n=2$. In the same way the lowest d-level is the $3d$. The same notation is carried over to levels of the alkali atoms, although the levels no longer coincide; there is thus no such thing as a $1p$-level or a $2d$-level.

§ 8. If these series of levels were only of importance for clearing up optical spectra, so much space would not have been devoted to them. But, as the reader will be aware, the above classification of levels turns out to be of the first importance for the understanding of atomic and molecular structure. Of considerably less importance are the details of the patterns and energies, which

* It will be shown in § 9 that when the "spin" of the electron is taken into account there are twice as many patterns.

will be briefly dealt with in this section. It is only in systems with one electron, the H atom, the He^+, Li^{++} ions, etc., that the patterns can be obtained exactly.

The volume of a spherical shell of radius r and thickness dr is $4\pi r^2 dr$, and $[R(r)]^2$ gives the density of the electron cloud in this shell. The probability of finding the electron at a distance from the nucleus between r and $r+dr$ is not $[R(r)]^2 dr$ but

$$4\pi r^2 [R(r)]^2 dr = 4\pi [F(r)]^2 dr \qquad \text{......(15).}$$

It is not surprising then that in (87) $F(r)$ is found to satisfy a simpler equation than $R(r)$ itself. The forms taken by both $F(r)$ and $R(r)$ are of interest (fig. 21). It is easy to see that when $l=0$, leaving $V_l = -\epsilon^2/r$, a particular solution of (14) is

$$F(r) = re^{-r/a} \qquad \text{......(16).}$$

For differentiating (16) twice with respect to r, we find

$$\frac{d^2F}{dr^2} = \left(\frac{1}{a^2} - \frac{2}{ar}\right) F \qquad \text{......(17),}$$

which is of the required form, provided that $\dfrac{2}{a} = \dfrac{8\pi^2\mu\epsilon^2}{h^2}$. The bracket changes sign at a spherical boundary given by $r=2a$, and this must be where $W=V$. An allowed value of W will therefore be given by $(W+\epsilon^2/2a)=0$. This is, in fact, the lowest level of the hydrogen atom:

$$W = -\frac{2\pi^2\mu\epsilon^4}{h^2} \qquad \text{......(18),}$$

which agrees with the value of the observed ionisation potential of the hydrogen atom, 13·53 electron-volts.

The allowed energies of the excited levels are found to be 1/4, 1/9, ..., and so on, of the ionisation potential, the general expression being

$$W_n = -\frac{2\pi^2\mu\epsilon^4}{h^2}\frac{1}{n^2} \qquad \text{......(19),}$$

where n is the principal quantum number. When n is large, the spacing between the levels is evidently small, and the series tends to the familiar limit. Expression (19) shows that in a Coulomb field the quantum number l plays no part in determining the allowed energies. The smallest amount of internal energy that a hydrogen atom can take up, i.e. the first resonance potential, is by (19) equal to 3/4 of the ionisation potential.

To compare the patterns belonging to these various levels with one another, consider in what regions we shall have $W > V_l$ for the different curves of fig. 18. For the *d*-levels the allowed region obviously occurs at larger values of *r* than for the lowest *p*-level, and still larger than for the 1*s*-level. The volume through which the electron cloud is spread out increases rapidly with the quantum number *n*, which agrees with our ideas of electrostatic energy.

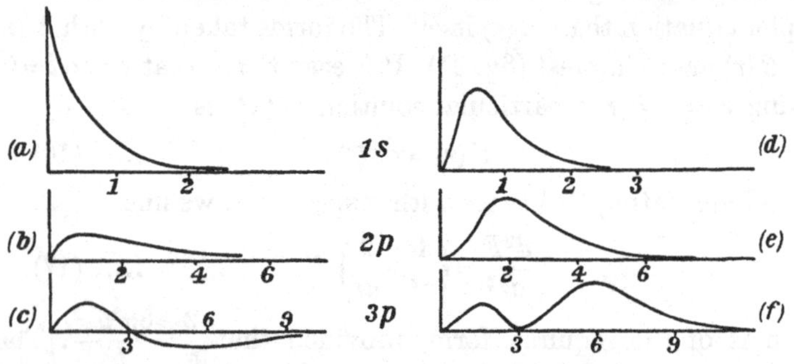

Fig. 21. In curves *a*, *b* and *c* the ordinates are proportional to $R(r)$, and in curves *d*, *e* and *f* to $[F(r)]^2$ for the same three levels, abscissae being *r* in Ångström units

In fig. 10 the patterns have been reduced for convenience all to the same size, but in fig. 21 the scale of each curve is given. The critical radius *a* which occurs in (17) has the value $5{\cdot}3 \times 10^{-9}$ cm., or about half an Ångström unit. Expression (16) shows that for the 1*s*-level the value of $F(r)$ is a maximum at a radius $r = a$; this maximum appears in curve *d* of fig. 21. For this level the expression for $R(r)$ with the required normalising factor inserted is

$$R(r) = \frac{1}{(\pi a)^{3/2}} e^{-r/a} \qquad \text{......(20).}$$

At $r = 0$ this has the value 5×10^{11} cm.$^{-3/2}$, mentioned in § 5; it is only at infinity that the value is zero. The difference in character between curves *b* and *c* and curve *a* is due to the fact that for all levels except *s*-levels there is a forbidden region enclosing the origin, with $W < V_l$. It is unnecessary to pursue this subject further; the curves for 3*d* and 4*d* resemble those for 2*p* and 3*p* respectively, but with larger values of r. When there are two or more maxima the outermost is always the largest as in curve f, indicating the presence of something analogous to the large orbit.

Finally, a word may be added about the variation of ψ with θ. The symmetry for $l=0$ has already been mentioned. When $l=1$, the variation is very simple, being $\cos\theta$ when combined with $m=0$, and $\frac{1}{\sqrt{2}}\sin\theta \begin{matrix}\cos\\ \sin\end{matrix}\phi$ when combined with $m=1$. When $\theta=90°$, $\cos\theta=0$ and gives rise to a nodal plane cutting the atom in halves, as can be seen for the $2p$-pattern in fig. 10. For $m=1$ the

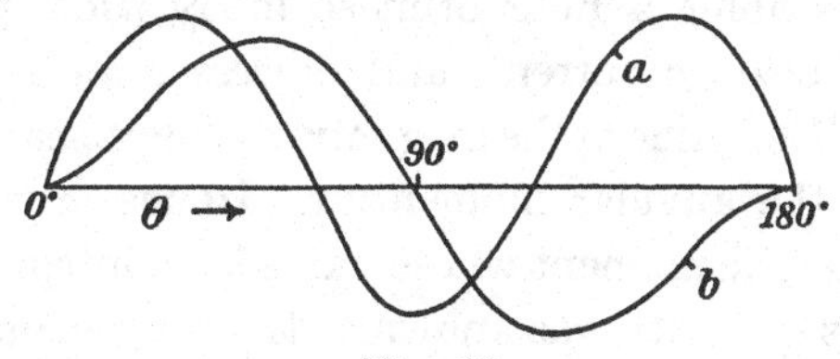

Fig. 22
Curve a is for $l=3$, $m=1$
Curve b is for $l=3$, $m=2$

same nodal plane occurs rotated through a right angle. For $l=2, 3$, and higher values, we have more complicated patterns, as in fig. 22.

For any level of principal quantum number n the pattern has always $(n-1)$ nodal surfaces. Take for example the $3p$-level. This is the second p-level and accordingly has one spherical nodal surface. There is also the nodal plane, mentioned above, passing through the nucleus, making two nodal surfaces in all. The same is true of all the levels.

As examples of the complete function $R(r).\Theta(\theta).\Phi(\phi)$ we may give the expressions for the $2s$- and $2p$-levels of the hydrogen atom. They are

$$\left.\begin{array}{lll} 2s & & b\left(2-\dfrac{r}{2a}\right)e^{-r/2a} \\ 2p & \left\{\begin{array}{l} m=0 \\ \\ m=1 \end{array}\right. & \begin{array}{l} \dfrac{br}{a}e^{-r/2a}\cos\theta \\ \dfrac{br}{a\sqrt{2}}e^{-r/2a}\sin\theta\,.\,e^{i\phi} \end{array} \end{array}\right\} \quad \ldots\ldots(21),$$

where
$$b=\frac{1}{4\sqrt{2\pi^3a^3}}.$$

§9. MAGNETIC MOMENT AND ELECTRON SPIN

The model of the hydrogen atom in its normal state, given by quantum mechanics, is so different from our previous conceptions that we are forced to make a revision of our ideas as to the magnetic properties of atoms, and in particular to a re-interpretation of the Stern-Gerlach experiment. The ideas of the older quantum theory may be briefly summarised as follows:

(1) One knows that any circuit carrying an electric current behaves as a magnet when placed in a magnetic field; this is the basis of the moving-coil galvanometer and loud-speaker. Now an electron describing a Bohr orbit so many times per second constitutes an electric current, and possesses an angular momentum $lh/2\pi$. The value of the magnetic moment bears a definite ratio $\epsilon/2\mu c$ to the angular momentum. Hence any quantised orbit has a magnetic moment whose value is an integral multiple of $\epsilon h/4\pi\mu c$. This quantity was known as the Bohr magneton, and has the value $9{\cdot}2 \times 10^{-21}$ e.m.u.

(2) It was found possible to account quantitatively for the simple Zeeman effect by assuming that when an atom was placed in a magnetic field, the plane of the orbit could take up only certain directions with respect to the field. For the normal unexcited hydrogen atom, for example, it was supposed that the magnetic axis must be either parallel to the field, or anti-parallel. In an assembly of such atoms half would set themselves one way, and half the other.

(3) Next consider a small magnet placed in an inhomogeneous field. If the north pole is in a stronger part of the field than the south pole, there will be a force acting on the magnet as a whole. If the south pole is in a stronger part of the field than the north, there will be a force acting in the opposite direction. If then the ideas in (2) as to the orientation of atomic magnets in a field were correct, it should be possible to separate one half of the atoms from the other half by passing them through an inhomogeneous field.

The experiment was successfully performed with hydrogen-like atoms and later with hydrogen atoms from a discharge tube. A stream of atoms was passed between the pole pieces of an electromagnet in a high vacuum. The faces of the pole pieces had been shaped to give as inhomogeneous a field as possible. When the magnetic field was turned on, the fine stream of atoms was found to divide into two streams with a narrow blank space between them, as had been anticipated. Nevertheless the original interpretation of this experiment must be rejected, especially as regards atoms whose normal state is an s-state.

In the first place the hydrogen atom no longer contains a flat electron orbit which can be oriented in the field. It must be pictured as a spherically symmetrical electron cloud, devoid of angular momentum or magnetic moment. And the same is true of all hydrogen-like atoms in their normal state. A different origin must therefore be found for the magnetic moment which is undoubtedly observed in the Stern-Gerlach experiment. The explanation is that the electron itself is a little magnet, and in this experiment we are measuring the magnetic moment of the electron itself; its value is that found in the Stern-Gerlach experiment and originally assigned to the orbit, i.e. one Bohr magneton. At the same time an angular momentum, known as the "spin" momentum and equal to $h/4\pi$, is now assigned to the electron. For atomic states other than s-states, i.e. when l is not zero, the resultant magnetic moment of the atom will be a vector sum of the spin moment of the electron itself and of the moment given by the value of l for the atom.

Another innovation is that quantum mechanics leads one to adopt a new language in which to describe magnetic forces. It will have been noticed that in this book the acceleration of particles is scarcely mentioned. In quantum mechanics attention is always directed to potential and total energies. The intensity of a field never enters the discussion except indirectly as the slope of a V-curve. It is quite in keeping with this point of view that one should confine attention to magnetic energies, asking no questions as to orientation by magnetic couples. If we place a hydrogen atom in a magnetic field, the shape of the atomic potential box is slightly modified. The ψ-pattern belonging to any initial allowed energy no longer fits, hence the allowed energy is shifted by an amount proportional to the field. The magnetic moment is the quantity which measures the change in energy. From the Stern-Gerlach experiment on hydrogen-like atoms we conclude that for an electron with spin in any field H there are two allowed energies, one higher and the other lower than the initial energy, the shift being $\pm MH$, where the magnetic moment M has the value of one magneton.

If the electrons themselves have a magnetic moment, it looks at first sight as if we ought to be able to perform a Stern-Gerlach experiment with a stream of free electrons, without using atoms as their carriers. But it turns out that the mass of the electron is so small that we cannot obtain a sufficiently well-defined beam of electrons, for reasons given in the next chapter. The electrons within an atom, however, moving round the nucleus, produce a magnetic field, and perform a kind of Stern-Gerlach experiment upon themselves. In this internal magnetic field the spin moment of the electron has a certain energy which is quantised and can take either of the two values $\pm MH$. For every set of values of the quantum numbers n, l, and m, we have now a pair of ψ-patterns belonging to energies which (except in the case of s-states) do not coincide. The levels thus occur in pairs, whose members may be distinguished by means of a fourth, or spin quantum number. In the case of s-levels, as we have seen, there is no angular motion associated with the ψ-pattern, and consequently no internal magnetic field to split the level. For s-states then the effect of spin is to give two possible states of the same energy. The separation of all other levels gives rise to spectrum lines of different frequencies; this may be taken roughly as the origin of the familiar twin sodium D lines.

This seems the most convenient point to mention the spin of the proton. When the spin of the electron was discovered, a spin angular momentum of $h/4\pi$ was assigned to the proton also, and this would give rise to a much smaller magnetic moment than that of the electron. The ratio between the two was again taken to be $\epsilon/\mu c$, where μ is the mass of the particle. Hence the magnetic spin moment of the proton should be about 1840 times smaller than that of the electron; see, however, § 3 of Chapter VII. The discussion of electron spin will be resumed in § 3 of the next chapter.

§ 10. SIMPLE PROBLEMS

From this long discussion of the hydrogen-like atom it will be seen that even an elementary treatment of the simplest atomic problem is somewhat complex. It is worth noting that the occurrence of several ψ-patterns belonging to one value of the energy is always found for potential boxes possessing symmetry. For example, compare an isotropic harmonic oscillator, whose potential energy is given by

$$V = \alpha(x^2 + y^2 + z^2),$$

with a harmonic oscillator, whose potential box is not spherical, i.e. with $V = \alpha x^2 + \beta y^2 + \gamma z^2$. The former possesses the same number of possible ψ-patterns as the latter, but many of these coincide as regards their energy; the allowed energy levels are, in fact, still given by expression (8), p. 24, but the quantum number n is now the sum of three independent quantum numbers n_x, n_y, n_z, each of which may take the values 0, 1, 2, ... To each combination of n_x, n_y, and n_z belongs a unique ψ-pattern. There will, for example, be six patterns belonging to the energy given by $n_x + n_y + n_z = 2$, namely

n_x	1	1	0	2	0	0
n_y	1	0	1	0	2	0
n_z	0	1	1	0	0	2

When n is large the number of patterns for a single energy level is approximately $n^2/2$. The harmonic oscillator whose potential box is not spherical possesses the same number of patterns but the energies do not coincide.

Fortunately, many atomic and molecular problems may be studied to some extent in one dimension, with results agreeing with those obtained by the proper analysis in three dimensions. For this purpose we need to follow up the methods introduced in § 4. In the following examples we shall consider only the lowest allowed level of each system. We shall often make use of the property of a sine curve, that where the ordinate is large the slope of the curve is small, and where the ordinate is small the curve is steep.

(1) Compare two potential boxes of the same width but of different depths, and contrast the ψ-patterns which will fit into them. It is clear that the wave-length which fits into the box with high walls (fig. 23*a*) is shorter than that which fits into the box with low walls (fig. 23*b*), since the latter needs a gentle slope at each boundary, to join on the exponentials with a small value of k. Now a shorter wave-length in the box means a larger momentum and a larger kinetic energy there; hence in the former case the

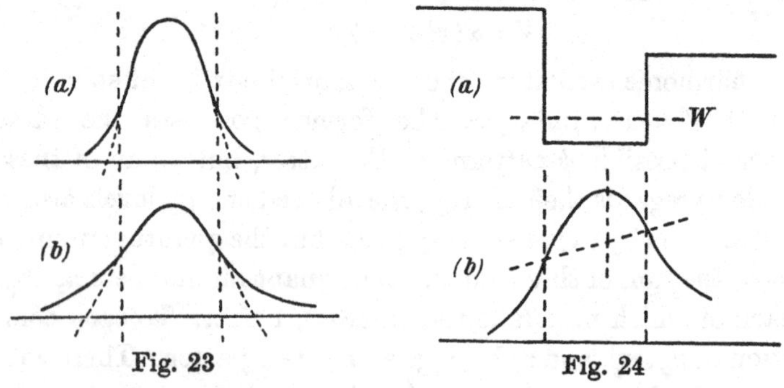

Fig. 23 Fig. 24

lowest allowed level will be further from the bottom of the box than in the latter.

(2) The form of the pattern for the lowest level of the box shown in fig. 24 follows at once. On the right, where the wall is low, the value of k will be less than on the left, where the value of $(V-W)$ is greater. Hence when the portion of sine curve is fitted in between, it must be put "out of centre", so that the steep part joins up smoothly on the right, as in fig. 24*b*; there is just one value of the energy W for which this is possible with no node.

(3) In $OPQ'T$, fig. 25, we have a potential box whose ψ-patterns we already know; but consider what will be the patterns for the V-curve $OPQRST$, where ST and PO are supposed to extend to infinity in either direction. In the region QR the value of $(V-W)$ is less than in the region beyond S. Hence in the expression e^{-kx}, k has a smaller value in the region QR than beyond;

let these values be k_1 and k_2, given by (5). We are of course not allowed to have a kink in the curve where e^{-k_1x} changes over to e^{-k_2x}, as in fig. 25*b*; how can this be avoided? The portion beyond S certainly cannot be tampered with, since ψ must fall to zero at infinity. A curve must therefore be found for the region QR which will join on smoothly; this can be obtained by using a mixture of increasing and decreasing exponentials in this region,

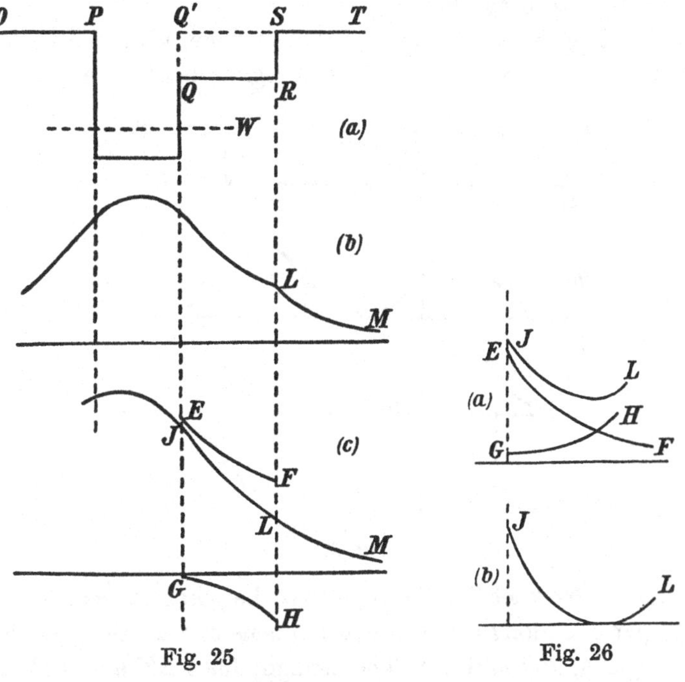

Fig. 25

Fig. 26

as shown in fig. 25*c*. Here the curve EF represents Ce^{-k_1x}, and the curve GH represents De^{k_1x} (D having a small negative value). Adding together the ordinates of EF and GH we obtain the curve JL, which evidently has a steeper slope at L than the curve EF has at F. By adjusting the values of C and D a curve can be obtained which joins on to the curve LM. In this way a curve can be obtained which falls smoothly to zero at infinity for any value of W, but only certain discrete values of W give curves which join on to the sine curve in the potential box.

(4) It is worth noticing that just as we obtained here a curve which became steeper than e^{-kx}, so we can obtain a curve which becomes less steep, or which even reverses its slope. This is shown in fig. 26*a*, where D is given a small positive value, and JL is again the sum of the curves EF and GH. If the value of D is still smaller, ψ may fall to a negligible value for quite a long distance, but will eventually rise again when De^{kx} becomes appreciable, as in fig. 26*b*.

(5) Looking back over what we have done, we recall that a potential box represents an allowed region in which a particle

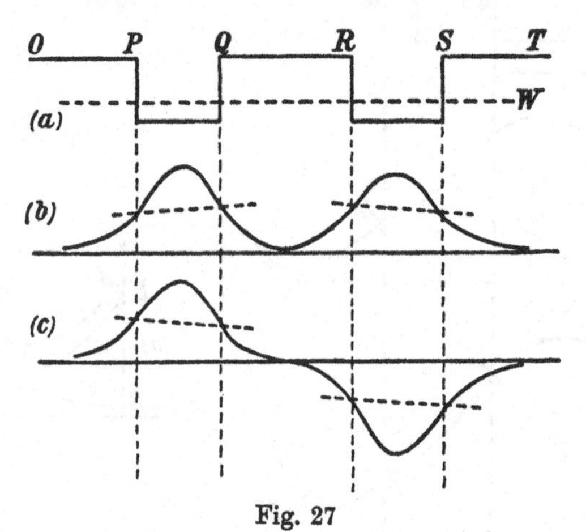

Fig. 27

may move. Now let fig. 27*a* represent the potential energy of a single particle along a line which happens to pass through two such regions, PQ and RS. The straight lines ST and PO are supposed to extend to infinity in either direction. In classical mechanics the two boxes would be quite independent. If a particle were placed in the potential box on the left, the existence of another box at a large distance from it could not possibly affect the motion of the particle. And if it were placed in the box on the right, the existence of the box on the left could not be relevant to its motion. In quantum mechanics, however, the ψ-pattern belonging to any allowed energy W extends through the whole of space. There will therefore no longer be a set of patterns for each box separately. Any allowed energy must be determined by the whole V-curve including both boxes. And each ψ-pattern must belong to the pair of boxes conjointly. Further it is clear that there will be a set of discrete allowed energies, as for a single box. From S and from P there must be exponential tails extending to infinity in either direction, and the restriction in making $d\psi/dx$ continuous operates as in § 4, so as to give quantisation.

In accordance with equation (7) there must be a portion of a sine curve inside each potential box. And it is evident that the type of curve mentioned in the preceding paragraph is exactly what is needed in the intervening region QR in order to connect

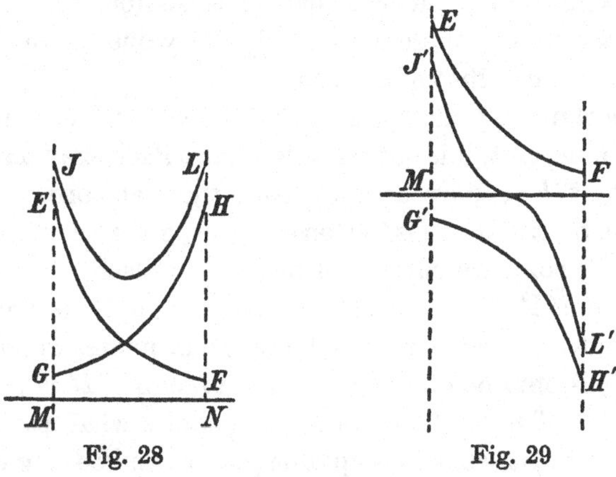

Fig. 28 Fig. 29

up the sine curves which we shall fit into PQ and RS. If the barrier QR is at all wide, the values of ψ^2 must fall almost to zero in this region, but must rise again in time to join on to the sine curve in the other box; for this the curve of fig. 26b is suitable. If the barrier QR is low and narrow, a curve like JL of fig. 26a is suitable. The problem is to adjust the values of C and D to give a curve which will join up smoothly everywhere.

A very important case in practice is where the boxes PQ and RS are identical (for example, potential boxes provided by two protons in the hydrogen molecular ion H_2^+). In this case the system will have complete symmetry about its mid-point. Hence the value of ψ^2 at any point in the right-hand half must be the same as at the corresponding point in the left half; in particular, ψ^2 must have the same value at R as at Q. How this can be achieved using $Ce^{-kx}+De^{kx}$ will be clear from fig. 28, where the points M and N correspond to Q and R. The curve GH, representing De^{kx}, is the mirror image of EF; hence when the ordinates of the curves are added together we obtain the symmetrical curve JL, as required for joining on to the sine curves. Let the distance $QR=d$, and at Q let $x=0$. Then at Q we have $\psi=C+D$, and the values of C and D must be adjusted so that

$$D=Ce^{-kd} \qquad \text{......(22)},$$

since $GM=FN=Ce^{-kd}$. A curve like JL is possible for any value of the energy W, but there will be only one value which gives a complete curve like that of fig. 27*b*.

The important fact now emerges that, for ψ^2 to have symmetry about the mid-point, it is not essential that ψ itself shall have the form of fig. 27*b*. ψ may have opposite signs at corresponding points, as in fig. 27*c*; for this too, on squaring, gives a symmetrical curve for ψ^2; compare curve *b* of fig. 15. In place of the curve JL we have $J'L'$ of fig. 29. Here the curve $G'H'$ is identical with EF but reversed; and on adding the ordinates of the two curves $J'L'$ is obtained. A curve of the type of $J'L'$ is possible for any value of W, and we are led to enquire what particular value of W will possess an acceptable pattern embodying such a curve. By looking at the details of the pattern we can decide whether it will belong to the same energy as the symmetrical pattern, to a higher energy, or to a lower energy. Comparing figs. 28 and 29, it will be seen that the slope of the curve $J'L'$ at J' is slightly steeper than the slope of EF at E, while the slope of JL at J is slightly less steep. These differences in slope evidently depend on the relative value of GM or $G'M$ to EM, and are very small when the barrier QR between the boxes is large.

When we come to complete the ψ-curve by joining a portion of sine curve to J or J', the problem is similar to that of fig. 23 above. We see that the sine curve to be fitted on to J must be joined slightly nearer its crest, while that for J' must be joined slightly lower down, with the final result that the value of ψ at Q is rather greater than at P (curve b, fig. 27) or else rather smaller than at P (curve c), as indicated by the slanting dotted lines in fig. 27. Into the box PQ we have to fit a rather smaller fraction of a wave-length than we should if the box RS were absent, or else (curve c) a rather larger fraction of a wave-length. Since the width of the potential boxes PQ and RS is fixed in advance, it follows that these two acceptable solutions of the Schroedinger equation have different de Broglie wave-lengths in the potential boxes; the λ to be fitted to the JL type is a little longer than that to be fitted to the $J'L'$ type. Now this difference in wave-length means a small difference in momentum and in kinetic energy. In other words we have a pair of energy levels near together. The symmetrical pattern of fig. 27b evidently belongs to the lower level of the pair, and the anti-symmetrical pattern of curve c to the upper. The energies of these twin levels lie on either side of the energy of the lowest state belonging to one box when the other is absent. The original ground level has become split into two by the presence of the second box in the neighbourhood. This will be true of each of the original levels. The original first excited level, whose pattern had one node in the box, splits into two levels whose patterns have the form of fig. 30. For the same reason as before the anti-symmetrical pattern again has the shorter de Broglie wave-length and thus belongs to the upper of the pair of levels.

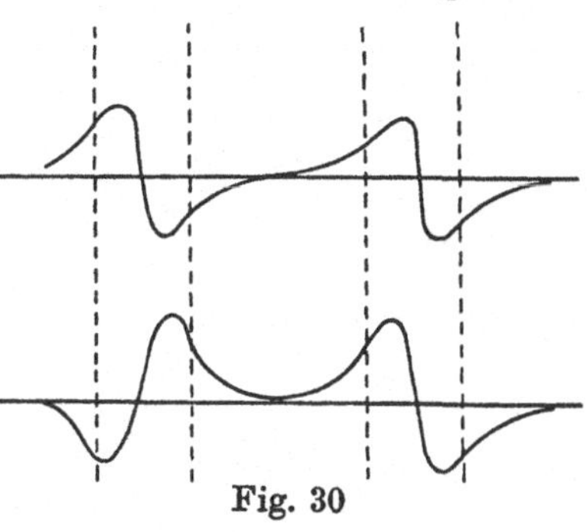

Fig. 30

In later chapters we shall have to discuss the interpretation of these patterns, and to make use of the important fact that twin levels diverge exponentially as the boxes are brought together, as in fig. 31, where the energies are plotted against the distance apart. It is clear from figs. 28 and 29 that the value of this divergence for any distance between the boxes depends upon the magnitude of FN compared with EM. That is to say, as the boxes are brought together, the divergence begins to be important at that distance d which makes the value of e^{-kd} appreciable. Although, for the sake of simplicity, rectangular potential boxes were taken in fig 27, all the main results are quite independent of the shape of the boxes; the anti-symmetrical pattern always belongs to the higher level of the pair.

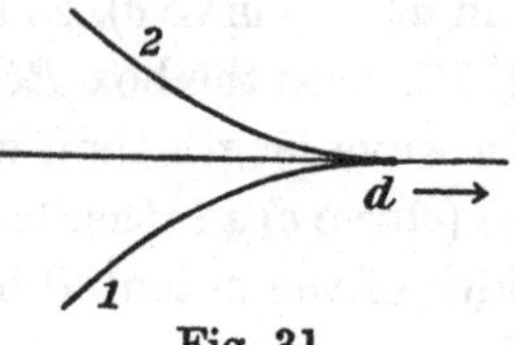

Fig. 31

CHAPTER IV

§1. THE UNCERTAINTY PRINCIPLE

To obtain a better understanding of ψ-patterns, one must look more closely into the factors which govern their form. The most delicate ideal observation which we can imagine is that in which we use only one incident quantum of light. In light of frequency ν each quantum moves with a momentum equal to $h\nu/c$, where h is Planck's constant and c is the velocity of light. It is the presence of this momentum which causes the recoil of any small particle with which the quantum collides. This disturbance, mentioned in Chapter II, is unavoidable, and we should not mind very much how large the recoil was, provided that we could make an exact allowance for it; if we could there would be no need to substitute patterns for definite values, for this is not due to our ignorance of the laws of the process. When a quantum is scattered through an angle θ by collision with a particle, the recoil momentum of the latter is given by (23) below; experimental studies of the Compton effect are in complete agreement with theory on this point. The only question is, how accurately can we measure the angle θ? This leads us to examine the conditions under which we may catch the scattered quantum in an ideal ultra-microscope.

We may attempt to measure the velocity of a particle by finding how far it travels in a very short interval of time. For this purpose we make two observations, using one quantum for each, incident along the x-axis, which is at right angles to the axis of the microscope, fig. 32. If the quantum is scattered through an angle θ the particle acquires a momentum whose x-component is

$$\frac{h\nu}{c}(1-\cos\theta) \qquad \text{......(23).}$$

The theory of the resolving power of the microscope, developed in the nineteenth century, supports the well-known fact that, in order to obtain good definition and resolution, one must use as large an aperture as possible; for the smallest distance that can be resolved is given by

$$\Delta x = \frac{c}{\nu \sin \alpha} \qquad \text{......(24)},$$

where α is the value of the aperture (the angle subtended by the objective lens) and ν the frequency of the light used for illumination. Now a large aperture is just what we wish to avoid. For in fig. 32*b*, where the aperture has been stopped down, any quantum of light which is observed to reach P is known to have come through the small hole in the diaphragm D, and hence θ is known with good accuracy. But in fig. 32*a* the quantum reaching P may have travelled along any path lying within the large cone of the aperture. Since $\cos \theta$ is not greater than $\sin \alpha$, the value of the recoil momentum may lie anywhere between $\frac{h\nu}{c}(1 - \sin \alpha)$ and $\frac{h\nu}{c}(1 + \sin \alpha)$. Hence the uncertainty in the momentum is

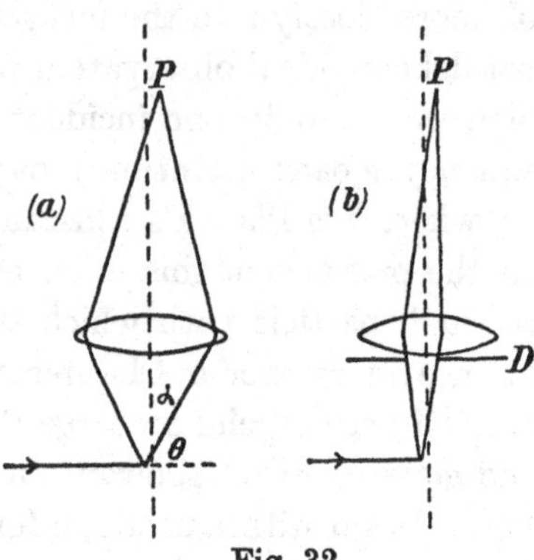

Fig. 32

$$\Delta p_x = \frac{2h\nu}{c} \sin \alpha \qquad \text{......(25)}.$$

This quantity is small only when $\nu \sin \alpha$ is small. Unfortunately by (24) when $\nu \sin \alpha$ is small the uncertainty in the position is large. We have then to compromise: the more precisely we measure the position, the less we know about the momentum, and the more precisely we measure the momentum, the less we know about the location of the particle. In fact, multiplying (24) by (25), we find that the product of the uncertainty in x and the uncertainty in p_x is of the order of Planck's constant

$$\Delta x \Delta p_x \sim h \qquad \text{......(26)}.$$

To get some idea of the magnitudes involved, we may evaluate the uncertainty in the position of an electron whose velocity is known within 20 per cent. to be 10^7 cm./sec. (corresponding to 1/40 of an electron-volt). The value of mv is about 10^{-20}, and thus Δp is 2×10^{-21}. Since $h = 6{\cdot}5 \times 10^{-27}$, we find that the uncertainty in position is more than 300 Ångströms. Conversely, if we knew the position of the electron to greater accuracy, this would involve a proportionately greater uncertainty in its velocity. For any macroscopic particle, on the other hand, the effect will not be noticeable. If, for example, the velocity of a mass of one milligram is known to be 1 cm. per second with an error of one in a million, the uncertainty in location is found to be negligibly small. This is because no massive particle suffers an appreciable recoil from incident radiation.

By imagining an analogous experiment, one can show that any measurement of an energy involves the measurement of a time, with a mutual limitation on the precision given by

$$\Delta W \Delta t \sim h \qquad \text{......(27).}$$

The period of vibration in a light wave is of the order of 10^{-15} sec.; if we wish to deal with times to this accuracy, it involves an uncertainty in the energy of more than one electron-volt. Expressions (26) and (27) give a quantitative measure to the ideas which were put forward at the beginning of Chapter II, as providing the *raison d'être* of ψ-patterns. Chapter III, on the other hand, has been spent in studying the solutions of equations (2) and (9)—equations which, as we have seen, grew out of the idea of de Broglie waves. We are led to look for connections between these various points of view. From the discussion of (26) we should expect the patterns giving the position of heavy particles to be less blurred than those of lighter particles, leading to classical behaviour in the case of massive particles. And in fact the values of e^{-2kx} calculated in the preceding chapter showed that the patterns derived from the Schroedinger equation have this property. In Note 3 of the Appendix on p. 150 it is shown that (26) may be regarded as equivalent to equation (2).

Attention has been concentrated in Chapter III on that type of ψ-pattern which gives expression to the uncertainty in the location of a particle. But we see from (26) that the uncertainty in its momentum is on exactly the same footing. It has already been mentioned that the momentum of a free particle must be dealt with by means of a ψ-pattern. For a particle in field-free space the relation $(W-V)=p^2/2m$ holds; hence corresponding to the pattern $\psi(p)$ there is a corresponding pattern $\psi(W)$ for the energy. And in general, when we have "prepared" any system in a state by means of preliminary observations, there must always exist a pattern such that $|\psi(W)|^2 dW$ gives the probability that the value of the energy lies between W and $W+dW$. This ψ must be expressed in ergs$^{-1/2}$, and the total area under the ψ^2 curve, fig. 33, must be made equal to unity, since the energy certainly has

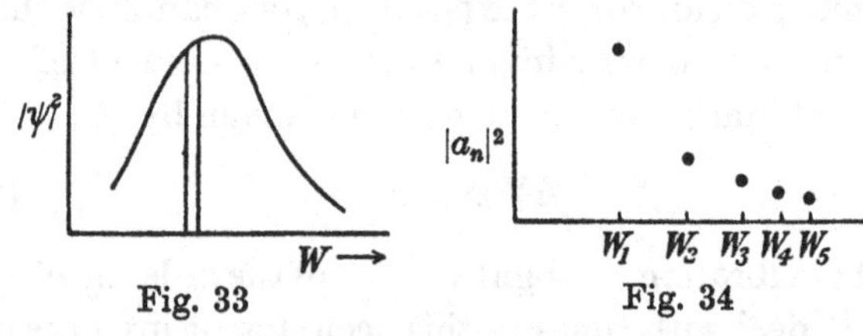

Fig. 33 Fig. 34

some value. Obviously such a curve applies only to free particles for which all energies are allowed. In this book we are more interested in particles bound together to give a quantised system which is stable only for certain discrete values of the energy, $W_1, W_2, W_3, \ldots$. For such a system the pattern for the energy distribution will not be a continuous curve, but a series of separate points (as in fig. 34), whose ordinates $a_1, a_2, a_3, \ldots$ are such that $|a_n|^2$ is the probability of finding the system in the nth state, with energy W_n. These a_n are just numbers, independent of x, y, z. Their values will depend on the way in which the system has been prepared, but must be such that

$$|a_1|^2+|a_2|^2+|a_3|^2+\ldots=1 \qquad \ldots\ldots(28).$$

§ 2. In the preceding chapter it was shown how those particular solutions of the Schroedinger equation which tend to zero at infinity yield at once the values of the characteristic levels of the hydrogen atom in agreement with observation. For this purpose it was not necessary to have a clear idea of the meaning of the solutions themselves. A caution was put forward against regarding each ψ-pattern as a substitute for the corresponding Bohr orbit. To see the reason for this, let us consider for example the ψ belonging to the energy W_2 of an atom, the energy of the so-called first excited state. Here, when ψ_2 has been properly normalised, $|\psi_2(x, y, z)|^2 dxdydz$ tells us what *would* be the probability of finding the electron in a small volume at the point x, y, z, if the energy of the system were *known* to be W_2 at the moment of observation—which would be only true if $|a_2|^2$ were equal to unity, and all the other a_n were zero. If there is uncertainty as to the energy of the system at the particular moment—and there usually is—if the atom may be in other states as well as state 2, then the probability of finding the electron in this volume $dx\,dy\,dz$ will depend on contributions from the ψ-patterns belonging to all the states concerned. And each contribution must obviously be weighted according to the probability of the state to which it belongs. The fact that the required ψ is the sum of all these contributions is expressed by writing

$$\psi(x, y, z) = a_1\psi_1 + a_2\psi_2 + a_3\psi_3 + \dots \qquad \dots\dots(29).$$

Every system must be described by means of a composite ψ of this kind except under special circumstances. To be able to add together different fractions of the various ψ-patterns in this way, one must have, so to speak, a "standard size" for each pattern. This is obtained by the normalisation, already described. The use of normalised patterns, in conjunction with (28), is sufficient to ensure that the composite shall be normalised.

The use of (29) can best be made clear by a simple example. Consider again the problem of fig. 27 on p. 44, in which we had two similar potential boxes at a considerable distance from one another. Let us prepare the system by putting a particle into one of the boxes, say the box on the left. It is reasonable to suppose that there will be a very high probability of finding the particle in the box on the left, if we look again after a short interval of time. Let us see whether the value of ψ^2 supports this expectation. If we square the ordinates of curve b of fig. 27, or those of curve c, we obtain in either case a curve like that of fig. 35*a*, in which the probability of finding the particle in the box on the right is just as great as that of finding it in the box on the left. This result, at variance with common sense, has only been obtained because we have made the mistake of using the ψ-pattern belonging to a single level to represent the system. We must at least take

$$\psi = a_1\psi_1 + a_2\psi_2 \qquad \text{......(30).}$$

If we put $a_1 = a_2$, and add the ordinates of curves *b* and *c* of fig. 27 before squaring, we obtain curve *b* of fig. 35, in which the particle is almost certainly in the box on the left. A composite ψ is clearly what is needed for describing the situation. In the same way, the pattern describing the situation when the particle is in the box on the right is obtained from fig. 27 by writing $a_2 = -a_1$, which gives curve *c* of fig. 35, and illustrates what is meant by saying that the values of the various a_n depend on the way in which the system has been prepared.

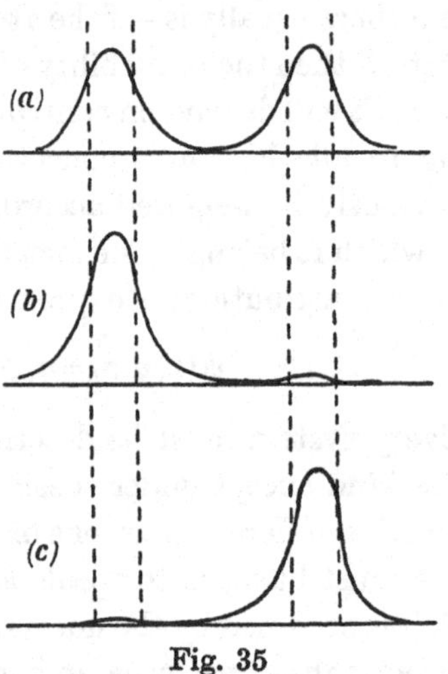

Fig. 35

These results point the way to a better insight into the meaning of the ψ-patterns which were obtained in Chapter III from the Schroedinger equation. Although it is often convenient to think of ψ-waves as existing in ordinary space like a real orbit, they are really devices which we use to represent our knowledge about the system. When we obtained curve *a* of fig. 35 by squaring either ψ_1 alone, or ψ_2 alone, we were *ipso facto* dealing with the case where the energy is known with an uncertainty smaller than the small separation between the twin levels; the contingent ignorance in the location of the particle is automatically expressed in fig. 35*a*. Conversely, if one knows which potential box the particle is in, one does not know which the energy state is. Thus the solutions of the Schroedinger equation (2), when used in the form of a composite ψ, exhibit a mutual uncertainty similar to that of (26).

If ψ_1 and ψ_2 are normalised, it is worth enquiring what must be the value of a in order that the pattern may remain normalised when, in forming a composite ψ, we write $\psi = a(\psi_1 + \psi_2)$. The area under the ψ^2 curve is given by

$$\int_{-\infty}^{\infty} \psi^2 dx = a^2 \left[\int_{-\infty}^{\infty} \psi_1^2 dx + \int_{-\infty}^{\infty} \psi_2^2 dx + 2\int_{-\infty}^{\infty} \psi_1 \psi_2 dx \right] \quad \ldots\ldots(31).$$

Of the three integrals on the right-hand side the first two are by the normalising condition equal to unity, and the third is easily seen to be equal to zero. Taking the curves b and c of fig. 27, it is clear that at any point in the right-hand half the value of $\psi_1 \psi_2$ is negative, while at the corresponding point in the left-hand half the product has an equal positive value; the value of this integral is therefore zero. It follows that the correct value of a in this case is $1/\sqrt{2}$. In (30) we thus have $a_1^2 + a_2^2 = 1$, in agreement with (28). The reader will see that by adding and subtracting the curves of fig. 30, as was done for those of fig. 27, patterns are obtained representing a particle in a higher level of either the box on the left or that on the right.

When we make use of a composite ψ to describe an atom, we are clearly attaching a new meaning to the phrase, if we speak of its "being in an excited state". According to (29) there is a probability of the atom having any of the energies W_2, W_3, ... as well as W_1. Since an excited state is unstable, the question arises as to whether we can ever prepare an atom with a definite energy. It may be pointed out that in every system of levels there is one which is unique, namely the lowest level of all. An atom in this stable level can leave it only by the acquisition of energy; so that, if neither radiation nor other particle is permitted to approach the atom, it will remain in this state indefinitely. In this case we are not compelled to use a composite ψ, but may have all the a_n except a_1 equal to zero. If we wish to study the behaviour of an atom with greater energy, with (say) energy equal to W_2, we write

$|a_2|^2=1$, and all the other $a_n=0$. (Strictly speaking, however, such a level, being unstable, cannot remain for a finite interval of time isolated from all lower levels of the system.)

§3. SYSTEMS CONTAINING MANY PARTICLES

In the problems so far considered we have always dealt with a potential box containing a single particle. In fitting patterns into the allowed region, we did not stop to consider whether the box could accommodate more than one particle; nor was there any need to do so. In the singly charged helium ion He^+ there is one electron and a vacancy for a second. And the ion Li^{++} has vacancies for two additional electrons; but these vacancies do not affect the levels or the patterns of the first electron bound in the box. We must now examine what happens when a box contains many particles. When to He^+ a second electron is added, each electron moves in a potential box provided jointly by the positive nucleus and the other electron. The discussion of hydrogen-like atoms in Chapter III made use of the idea of treating a single valence electron as moving in a potential box provided by the nucleus and the core electrons. The core of a hydrogen-like atom is almost the same whether the valence electron is present or absent; this is because the valence electron spends most of its time outside the core. This is an exceptionally simple case; usually the patterns are determined by mutual adjustment. For example, in the helium atom the form of the electron cloud representing one electron determines the shape of the potential box in which the other electron moves, and hence the form of its ψ-pattern; the latter in turn determines the shape of the box in which the first electron moves. (Another method of treating this type of problem will be given in Chapter VI.) In an atom containing N electrons we may regard each electron as moving in a potential box provided by the positive nucleus and $(N-1)$ electrons.

This idea may be extended to the free electrons in a metal. If a valence electron is free to move through the metal, and is only turned back at the surfaces, the whole piece of metal forms a huge potential box. Consider a piece of metal containing N monovalent atoms. Fixing attention on one free electron, we may consider it to be moving in a box provided by N positive cores and $(N-1)$ electrons. Every other valence electron is doing the same, namely moving in a potential box whose boundaries coincide with the surfaces of the metal. The possible ψ-patterns are the same for each electron, and are to be obtained as in Chapter III. Each ψ-pattern fills the whole volume of the metal uniformly, as in fig. 12, and dies away exponentially outside the surface. The same rules apply to a metal as to an atom with many electrons. But the problem of the metal is somewhat clearer, because the ψ-patterns belonging to high energies are distributed through the same volume as those of low energies, whereas in an atom the ψ-patterns for low energies are concentrated close round the nucleus.

In discussing a potential box containing many electrons, we may keep in view both the atomic and the metallic problems. The total energy is the sum of the energies of the individual electrons in their quantised levels. At absolute zero temperature the assembly of electrons in the box will settle down into the lowest state of which it is capable. The first part of the problem is to decide what is this state of lowest energy. In his theory of atomic structure Bohr long ago suggested that when electrons are introduced into an atomic box they occupy various levels, starting from the bottom. If one could start with a bare atomic nucleus with a large positive charge, and could build up an atom by adding electrons one by one, what would happen? The first two electrons, Bohr suggested, would go into the lowest level, the K-level, the next eight electrons into higher levels, the L-levels, the next eighteen electrons into still higher levels, and so on; even at zero temperature the electrons could not sink down into lower levels than these. This scheme was merely an empirical idea introduced to account for X-ray levels and for the periodic table of the elements, in which it was very successful. When in 1926 the idea of the spinning electron had been introduced, together with the rule that the spin quantum number could take two values and no more, it was possible to express Bohr's scheme by means of the rule: In an atom no two electrons can be in identical quantum states. In the K-level, for example, there is room for a pair of electrons with opposite spins, but there is no room for a third electron. The rule prevents electrons from crowding into the same level, and is known as the Exclusion Principle. It applies to any potential box, small or large. Even in a piece of metal no two electrons can be in exactly the same quantum state. The free electrons are therefore distributed through millions of levels, only a small fraction of them being accommodated in levels of low energy. The Exclusion Principle will be applied to atomic structure in § 4, and to the free electrons of metals in § 5.

§ 4. ELECTRON CONFIGURATIONS AND THE PERIODIC TABLE

With the ψ-patterns of Chapter III at our disposal, let us reconsider Bohr's idea of building up a heavy atom. Starting with a bare nucleus with a large positive charge $Z\epsilon$, let us bring up the first electron. The allowed levels and the possible patterns for this one electron are obtained at once from those of hydrogen; only alterations in the constants are required, since the potential energy occurring in the Schroedinger equation will be $-Z\epsilon^2/r$ in place of $-\epsilon^2/r$. The scheme of energy levels is the same as that given by (19), only each level is exactly Z^2 times as deep. The ψ-patterns are an exact replica in miniature of the hydrogen ψ-patterns, with the radius reduced exactly Z times in scale. We are here only interested in the ground level into which this electron will go; the negative energy of this level will be exactly Z^2 times the ionisation potential of the hydrogen atom (13·53 electron-volts). For example, when $Z=80$ the value will be 86,400 ϵ-volts. This is the work required to remove this electron from the nucleus when no other electrons are present. When the 79 other electrons have been added to build up a complete mercury atom (with atomic number 80), the work required to remove this electron, now a K-electron, will be somewhat less; but not very much less, for the value is known to be 82,900 ϵ-volts (the value obtained from the wave-length of the K-absorption edge for X-rays). Since the same is true for other elements, the energy of the K-level is nearly proportional to the square of the atomic number Z; so that if the square root of the energy is plotted against Z, we have a straight line, the well-known Moseley diagram.

Going on now to the addition of further electrons to these K-electrons we see that the application of the Exclusion Principle means that the levels occupied by the various electrons will follow the same classification as the possible excited levels for the single electron of the hydrogen atom, namely, $1s$, $2s$, $2p$, $3s$, $3p$, $3d$, etc. The principal quantum number no longer gives the value of the energy by (19), but it still specifies the number of spherical nodal surfaces in the electron cloud, namely $n-1$ for s-levels, $n-2$ for each p-level, $n-3$ for each d-level, and so on. The values of the energies and the forms of the patterns are determined by the mutual adjustment of the electrons among themselves, and can only be determined by laborious calculation. A knowledge of the types of the levels, however, enables us to account for the principal physical and chemical properties of the elements in the periodic table.

The two K-electrons, being both s-electrons, co-operate with the nucleus in providing a spherically symmetrical positive core. The lowest vacant level is a $2s$, with a pattern whose principal density lies outside this core. From electrostatic principles ψ-patterns whose main density is relatively distant from the nucleus should belong to high energies; and they do. The first ionisation potential of lithium is less than half that of hydrogen, as will be seen from the Table given at the end of this book.

There is evidently room in the L-shell for eight electrons, i.e. a pair of $2s$-electrons with opposite spins, and three pairs of $2p$-electrons. The quantum numbers of each of the four pairs are

n:	2	2	2	2
l:	0	1	1	1
m:	0	1	0	-1

The next electron must go into a level with $n=3$ outside the core containing ten electrons. Consequently the element with atomic number 11 is monovalent sodium, resembling lithium. In discussing figs. 18 and 19, it has already been mentioned that for each hydrogen-like atom the normal lowest level of the valence electron is an *s*-level. It is in fact a 3*s*-level for sodium, a 4*s*-level for both potassium and copper, and so on, as will be seen from the Table at the end of the book. In stating the structure of an atom, the number of electrons of a particular kind is usually expressed by an index. In the standard notation the structure of the nitrogen atom in its normal state is written $1s^2 . 2s^2 . 2p^3$, meaning two 1*s*-electrons, two 2*s*- and three 2*p*-electrons. In the same way the structure of sodium is

$$1s^2 . 2s^2 . 2p^6 . 3s.$$

The *M*-shell can accommodate eighteen electrons, that is, nine pairs whose quantum numbers are

l:	0	1	1	1	2	2	2	2	2
m:	0	1	0	−1	2	1	0	−1	−2

This accounts for the eighteen elements in the table between neon 10 and copper 29, which has the *M*-shell full and one electron in the *N*-shell. It will be noticed that of these eighteen electrons twelve have values of m differing from zero. But the magnetic moments show no tendency to pile up and to give large values in the heavier elements; for it will be seen that for every pair of electrons with $m=1$ there is a pair with $m=-1$, and for every pair with $m=2$ there is a pair with $m=-2$, which cancels out their magnetic moment. In the same way the spin moments of the individual electrons show no tendency to pile up. For in any closed shell they occur in pairs such that in a magnetic field one acquires energy and the other loses an equal amount of energy; by definition the magnetic moment of the pair is zero. In the heaviest elements there are more than forty such pairs making no contribution to the magnetic moment of the atom.

Coming finally to the electron cloud itself, we find an interesting result for the contributions of *p*-electrons. As mentioned on p. 37, when $l=1$ the variation of Θ is given by $\cos\theta$ when $m=0$, and by $\frac{1}{\sqrt{2}}\sin\theta$ when $m=\pm 1$. The contribution made to the electron cloud of the atom by any p-electron therefore varies as $\cos^2\theta$ or as $\frac{1}{2}\sin^2\theta$. If in any atom there are three similar p-electrons with values of m equal to 1, 0, and -1, their combined contribution to the atomic electron cloud will be spherically symmetrical, since by (12) $|\Phi|^2$ is constant, and further

$$\frac{1}{2}\sin^2\theta+\cos^2\theta+\frac{1}{2}\sin^2\theta=1 \qquad \text{......(32).}$$

In the tables of quantum numbers given above for the L- and M-shells, it will be seen that there are three pairs of such similar p-electrons. The spherical symmetry of the complete L-shell follows at once. In the M-shell there are in addition five pairs of d-electrons. For five similar d-electrons a relation similar to (32) holds, which leads to the complete spherical symmetry of the M-shell. In the older quantum theory the readiness of a halogen atom to take up an electron had been correctly ascribed to the tendency of the incomplete shell to take up the one missing electron. Of this quantum mechanics gives now a more detailed picture. In the electron cloud of a halogen atom there is a kind of "hole" whose shape is that of the pattern belonging to the quantum numbers of the missing electron. When an extra electron is taken into the shell, this hole is filled up, giving complete spherical symmetry.

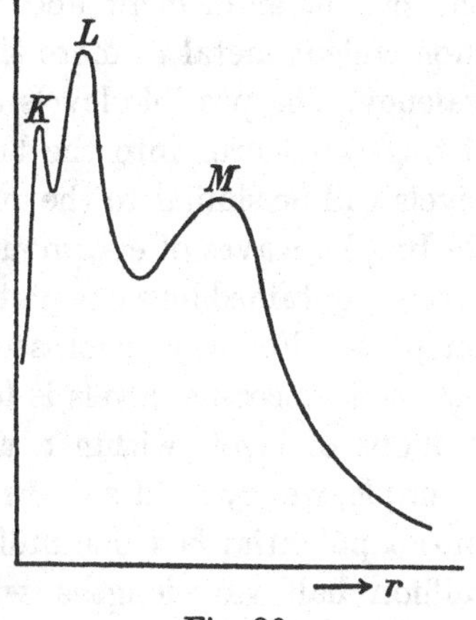

Fig. 36

The electron clouds representing the various K-, L-, M-shells will of course overlap, and there remains the resultant cloud to be considered, i.e. the radial factor of the density of the complete atom with its mutually interfering electrons. Before the introduction of quantum mechanics it was recognised that this distribution of density was very important as governing the scattering of X-rays by atoms of different elements. In those days such a curve as fig. 36, where the electron density is plotted against the radius, would have been regarded as giving the time-average of well-defined electron orbits of various shapes overlapping. Now, however, we regard a ψ-pattern as the only legitimate way of describing the electrons in an atom.

§ 5. METALS

It is often said that in the absence of a current no electric field exists inside a conductor. If this were literally true, the potential energy of an electron would be the same everywhere throughout the interior, and the potential box would be like fig. 5. In reality the positive core of every atom in the metal gives rise to a local intense field, not unlike fig. 7. But many of the properties of metals may be deduced by imagining these positive charges to be smoothed out into a uniform cloud, which is almost neutralised by a similar uniform negative cloud due to the $(N-1)$ electrons. That is, the potential box for any valence electron is to be of the simple type of fig. 5, with a depth that is characteristic of the particular element and varies from about 6 ϵ-volts for a monovalent metal to more than 12 ϵ-volts for metals of higher valency. The possible levels are found by the usual method of fitting ψ-patterns into the box. Those belonging to the lowest levels will be similar to the curves of figs. 12 and 15, but will be de Broglie waves of enormous length. The spacing between the levels is obtained by a calculation like that given for atomic levels on p. 24. For any macroscopic piece of crystal the intervals between successive levels is found to be so small that there are millions of levels within a range of one electron-volt. If, for example, we try to fit a de Broglie wave-length of 10 Ångströms into a potential box one millimetre in length, we see that two million half-wave-lengths will fit in; this is therefore the 2,000,000th level of the set. We see further that the larger the piece of metal, the closer will be the spacing of the levels. When the problem is solved for a cube, there are for each allowed energy a number of possible patterns, whose energies coincide; for the nth level there are approximately $n^2/2$, as in the problem on p. 41. For a piece of metal of the same size which is not a cube, there will be the same number of levels, but they will fail to coincide. The scheme of levels which we obtain from this simplified model of a metal is very similar to that at which one arrives from taking properly into account the periodic structure of metallic crystals, which will be considered in Chapter VIII.

To find what will be the state of the metal at absolute zero temperature, electrons have now to be inserted into these levels in accordance with the Exclusion Principle. Each level can accommodate a pair of electrons with opposite spin. Com-

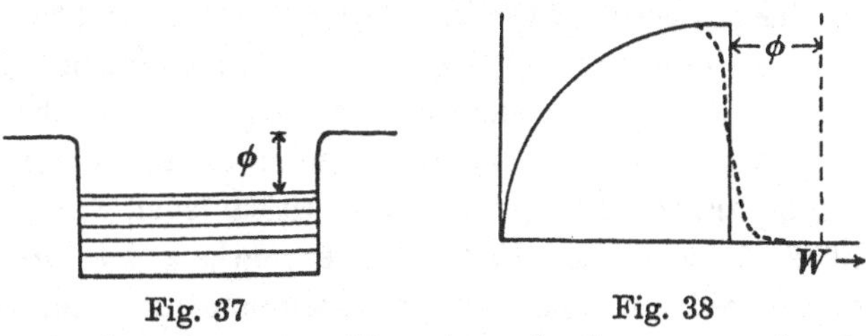

Fig. 37 Fig. 38

paratively few electrons will go into the low energy levels; but the number increases as n^2. At the millionth level, for example, there will be vacancies for about 10^{12} electrons. In this way, although there may be more than 10^{22} free electrons per cubic centimetre, we can find room for them all without violating the principle that no two electrons may be in the same quantum state; there are, in fact, always plenty of levels to spare. We will consider first a typical metal at zero temperature, and can afterwards add the thermal energies for higher temperatures. The depth of the potential box is supposed to vary from about 6 to 18 ϵ-volts for different elements, and the electrons fill up about the lower two-thirds of this, leaving the rest of the box, above a certain level, completely empty. This empty range, fig. 37, provides the characteristic work function ϕ of the metal. The resemblance of this quantity to a latent heat of vaporisation was described in Chapter I. For different elements the observed value of ϕ ranges from 2 ϵ-volts for the alkalis to more than 6 ϵ-volts for platinum. For any particular element the value of ϕ is independent of the size of the piece of metal considered; for, though in a larger piece there are more electrons to put into the box, yet this is exactly counterbalanced by the fact, mentioned above, that the spacing of the levels is closer in a larger potential box. The range of energies over which the electrons are spread depends only on their density, not on their total number. Since the spacing of the levels is so close, we may speak of them as if they formed a continuous instead of a discrete set; if the number of electrons per cubic centimetre having energies between W and $W+dW$ is plotted against W, the full curve of fig. 38 is obtained. In classical mechanics one supposed that at very low temperatures the kinetic energy of the free electrons would tend towards zero like the kinetic energy of molecules of a gas. But in figs. 37 and 38 the average kinetic energy is very high even at absolute zero. When the temperature is raised, some electrons from the highest levels are thrown up above the critical level, which may be called W_0, so that the latter no longer marks a clear division between occupied and vacant levels. At temperature T the levels at the critical level are half full, and above them the number of electrons falls off as $e^{-(W-W_0)/kT}$, as shown by the dotted line in fig. 38. At room temperature the fall of this exponential is so sudden that the curve would be indistinguishable from the straight line in fig. 38. The majority of the electrons are still in the same levels as they were in at zero temperature. They therefore make a much smaller contribution to the specific heat of the metal than free electrons would do in classical theory. Whether one regards the division between the occupied and vacant levels as still fairly sharp depends upon the physical process under consideration. In the photoelectric effect the division is sufficiently sudden to give the sharp threshold frequency, observed even above room temperature, when for the incident light $h\nu=\phi$. In thermionics, on the other hand, it is the electrons which are thrown up above the critical level by collisions within the metal which are responsible for the whole effect, even in dull emitters; this will be discussed in Chapter VIII.

Any gas exerts a pressure on its boundaries and tends to expand. If, then, there is some truth in the above method in which the metallic electrons are treated as a monatomic gas, we ought to offer some explanation of the factors which determine the specific volume of each element. When we diminish the volume of a piece of metal by compression or enlarge it by stretching, in either case we do work. If, then, the energy is plotted against the volume, the curve must have a minimum at the observed specific volume, like curve C in fig. 39. Comparing this curve with the minima in figs. 18 and 51, we might suspect that it is the resultant of two curves such as curves A and B, one representing an energy of attraction, the other an energy of repulsion.

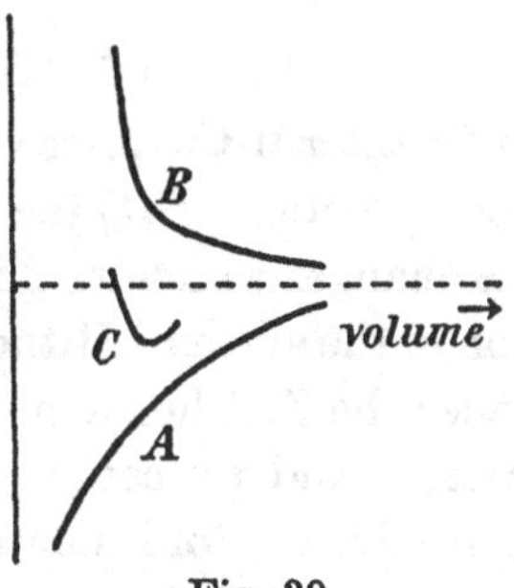

Fig. 39

When we diminish the volume of a piece of metal, we are diminishing the volume of the potential box in which the electrons move. And since each allowed pattern has a wave-length λ which must fit into the box, it follows that every wave-length must be proportionately diminished when we compress the metal. This applies to all the millions of levels. But a shorter de Broglie wave-length means a higher kinetic energy for the electron; the kinetic energy of every free electron has been increased. At the same time, when we compress the metal the average distance between the positive and negative charges is diminished, so that the potential energy is lowered to some extent. According to this simple model, if the electrostatic energy is given by curve A, and the total kinetic energy of the electrons by curve B, the sum of the ordinates, curve C, represents the total energy. The slope of curve B at any point represents the pressure of the electron gas.

This brief discussion of fig. 37 has been included here only as an introduction to the study of metals which will be given in Chapter VIII, taking into account the fact that they are crystalline.

CHAPTER V

§ 1. THE MOVEMENT OF PARTICLES

In Chapter II the aims of quantum mechanics were taken to be the prediction of (1) the possible ψ-patterns of a system, and (2) the changes which would take place when we make some alteration in the system. Hitherto we have been dealing with problems under the first heading—the permanent structure of atoms and metals. And we could go on now to obtain the ψ-patterns for molecules and for insulators by an extension of the same methods. It seems better, however, to postpone that work, and to devote this chapter to a preliminary survey of the mechanics which falls under the second heading—describing the motion of particles in physical changes.

A disturbance comes to an atom or molecule mostly in the form of an electric or electromagnetic field, which has usually been thought of as accelerating and deflecting the electrons and atoms. In quantum mechanics, however, the disturbance will always be considered as a change in the potential energy V. In the presence of the new added field the potential energy of any charged particle at a point x, y, z will be different from what it was initially; that is to say, there is an alteration in the V-curve. The original ψ-patterns are of course no longer appropriate to this modified V-curve; they are replaced by patterns of different shape and this represents a physical event.

We may conveniently distinguish two types of disturbance. In the first type (1) the disturbance is transient, and after a short interval the V-curve returns to its original form. Consider for example a hydrogen atom near which a fast electron passes. Initially the mutual potential energy of the atomic electron in the field of its nucleus has the form of fig. 7. As the fast particle goes by it superimposes on this an intense field, thus modifying the V-curve for perhaps 10^{-14} of a second, after which the potential energy near the nucleus is once more that of the simple Coulomb potential. When the V-curve has returned to its initial form, the original ψ-patterns are once more the appropriate ones—in fact, the only possible ones. But this does not mean that nothing has happened. For the state of the system is given by

$$\psi = a_1\psi_1 + a_2\psi_2 + a_3\psi_3 + \dots$$

and the various ψ's may now be mixed in different proportions according to new values of $a_1, a_2, a_3, \dots$. If the atom or molecule was initially in its ground level, it may have been raised to any one of its excited states. The solutions $\psi_1, \psi_2, \dots$ of equation (9) for a particular V are obviously unchangeable, and form a kind of alphabet of quantum mechanics; a composite ψ, formed from them, changes by a rearrangement of the proportions in which the component patterns are mixed.

(2) In the second type of disturbance the alteration in the potential energy is not transient. With an altered V-curve the original permanent ψ-patterns no longer give us correct information about the system. While things are in a state of flux, the probability $P(q)\,dq$ that any observable quantity has a value lying between q and $q+dq$ is no longer constant, but varies with the time. We need a method of finding the value of dP/dt. Since we always find P in the form $|\psi|^2$, it is natural to proceed by taking d/dt of ψ itself. The quantitative description of these two types of physical process is a very large subject, and will only be dealt with in Chapters IX and X. Here we shall merely clear the ground by considering what new principles must be introduced to describe the motion of particles. Obviously equation (9), which does not contain the time, can tell us nothing about movement; we need a new equation.

§ 2. Although in the last two chapters it has been possible to make good progress, only now are we confronted by the fundamental problems of quantum theory. For any method we may devise for dealing with disturbed atoms must be competent to deal with that most important type of disturbance, the absorption of light. And it was just on this absorption process that the divergence between classical theory and the older quantum theory was most unsatisfactory. According to classical theory the absorption of radiation by atoms and molecules was a resonance process, similar to that in acoustics. Electrons in atoms were supposed to move with characteristic frequencies which responded to the same frequencies in the incident light, thus picking out the narrow absorption lines from any continuous spectrum. Similarly, the atoms in molecules were supposed to move to and fro with characteristic frequencies which gave rise to absorption lines in the infra-red. These simple conceptions were undermined when the idea was introduced that radiation of any frequency ν could only be absorbed as a definite quantum of energy $h\nu$. To account for absorption one had now to provide not a resonance mechanism, but a mechanism for the absorption of definite amounts of energy, without worrying very much about the frequency. This was done by relating the frequency to the energy states W_m, W_n of the atom or molecule *before* and *after* the absorption, which is a very different thing from relating it to the frequency of a particle vibrating *during* the absorption process. It is one of the triumphs of quantum mechanics that it has succeeded in reconciling these two points of view.

To the ψ-patterns which were found to give us quantised levels must now be welded the idea of characteristic vibrations. In reference to fig. 15 attention was drawn to the resemblance of those ψ-curves to the forms taken up by the standing waves on a vibrating wire. The resemblance may be made closer if we suppose that at every point in a ψ-pattern the value of ψ, like the displacement of a wire, continually changes sign from positive to negative ν times per second. Proceeding by guess-work, we suppose that the frequency is related to the energy W by $\nu = W/h$, as for a Planck oscillator. Retaining for the symbol ψ the meaning which it has had all along, it is customary to use a capital letter, Ψ, for the product of ψ and its time factor ($\cos 2\pi\nu t - i \sin 2\pi\nu t$), where $i = \sqrt{-1}$ (see Note 2 of the Appendix, p. 149). Thus for the value of Ψ at any time t we write

$$\Psi = \psi e^{-2\pi i\nu t} = \psi e^{-2\pi i(W/h)t}$$

and for the conjugate complex

$$\Psi^{\star} = \psi^{\star} e^{2\pi i\nu t} \qquad \ldots\ldots(33).$$

Each pattern belonging to an allowed energy W_n thus vibrates with its own frequency W_n/h. This development has no effect at all on any interpretations of the pattern that have been made so far, since $\Psi\Psi^* = \psi\psi^* = |\psi|^2$. It will be understood that there is no reason why the ψ-curves of figs. 13, 15 and 27 should be drawn in those forms in preference to those of fig. 40. We now consider every Ψ to be oscillating between the two forms.

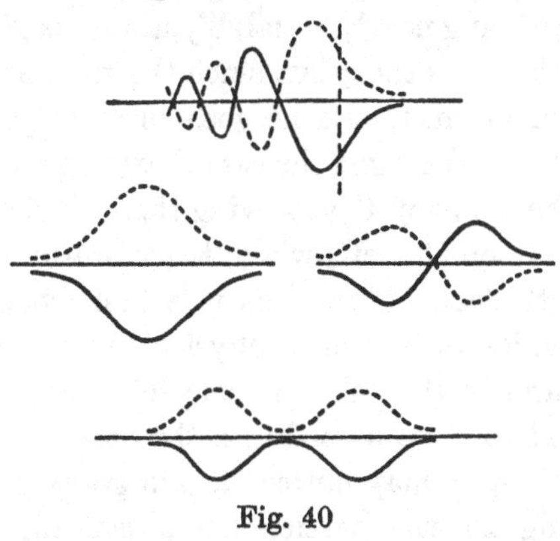

Fig. 40

§ 3. It is worth pointing out that, without making any further development, the vibrations of Ψ receive a neat application in the problem of fig. 27, p. 44. A particle put into the potential box PQ of fig. 27 will move to and fro, and will not always be turned back near P and near Q, but will sometimes penetrate a considerable distance beyond the boundary, the relative probabilities being given by the calculation on p. 19. We can hardly escape the conclusion that sooner or later in one of these excursions it will reach R, after which it will be found for a time in the box RS. There is of course no certainty in this, only a probability, which should be represented by a leaking of the probability pattern. Let us see whether these expectations are fulfilled without introducing any new assumption. The situation when the particle is known to be in the box on the left was obtained in fig. 35, p. 54, by adding together equal quantities of the twin ψ-patterns of fig. 27 b and c, which belong to the twin energies W_1 and W_2. It will be recalled that when this is done the patterns almost completely cancel each other out in the box on the right. Owing to the very small separation between W_1 and W_2 the vibration frequencies which we ascribe to Ψ_1 and Ψ_2 will be very nearly but not quite the same, namely $\nu_1 = W_1/h$ and $\nu_2 = W_2/h$. The state of the system is then to be represented by

$$\Psi(x,t) = \Psi_1 + \Psi_2 = a_1\psi_1 e^{-2\pi i\nu_1 t} + a_2\psi_2 e^{-2\pi i\nu_2 t} \quad \ldots(34).$$

It is obvious now that something is going to happen. The frequencies ν_1 and ν_2 being nearly equal, Ψ_1 and Ψ_2 begin by vibrating in phase with each other; but since the frequencies are not identical, the synchronism as time goes on must get worse and worse. In the box on the right the cancellation gets less and less complete, and the value of Ψ is growing there all the time at the expense of the box on the left, where the vibrations are getting out of phase with each other. This is a first example of how quantum mechanics describes a physical event. There is an increasing probability that the particle initially placed in one box will be found to have passed into the other.

At the same time we may notice that in going from one box to the other in fig. 27, the particle has passed right through a region where classically it could not go. This power of particles of burrowing through a potential barrier where $W < V$ plays an important part in quantum mechanics, and is known as the "tunnel effect". In the above problem we should anticipate that the larger the intervening barrier between the potential boxes the smaller would be the chance of the particle having slipped through it in a given short interval of time. We see at once that this expectation is realised; for the rate of leak given by (34) is proportional to the "beat frequency" of the component vibrations, namely $(\nu_2 - \nu_1)$, and hence to the energy separation $(W_2 - W_1)$; and the latter is proportional to e^{-kd}, where d is the distance between the boxes. Hence the initial rate of leak depends exponentially upon the width of the intervening barrier, as we should expect, and becomes negligible at long distances. When, on the other hand, the barrier is extremely narrow, it ceases to present an obstacle to the particle, so that the pair of boxes almost forms a single potential box.

§ 4. While the problem just discussed has illustrated an important rôle which the vibrations play in quantum theory, we may return now to the question raised in § 2 as to the possible correspondence of the frequencies to the characteristic frequencies of classical radiation theory, in spite of the fact that the patterns themselves belong to quantised levels. Consider the atoms in a flame, a stellar atmosphere, or a discharge tube which is emitting light; in the language of the quantum theory a certain fraction of these atoms is at any moment in each of the excited states. Similarly in a molecular gas, even at room temperature, molecules have various amounts of quantised rotational energy, as explained on p. 7, i.e. only a fraction of them are in their ground state. And it is not the same molecules which remain permanently in the ground state, but the molecules are continually going from one state to another by the absorption and emission of energy. Though it would seem at first sight as if some molecules should be represented by Ψ_1 belonging to the ground level, and others by Ψ_2, Ψ_3 and so on, yet we have to admit that in such an assembly we have no means of telling in which state any particular atom or molecule is at any moment. There are only certain probabilities of finding a molecule in state 1, 2, or 3, ...; and since these probabilities are the same for every molecule, we must not use different Ψ-patterns to represent molecules in different states, but a single pattern, the same for every molecule. This will be composite, as in (29):

$$\Psi = a_1\Psi_1 + a_2\Psi_2 + a_3\Psi_3 + \ldots \qquad \ldots\ldots(35).$$

For simplicity let us take first only two states, for which Ψ is given by (34), and its conjugate complex by

$$\Psi^\star = a_1^\star\psi_1^\star e^{2\pi i(W_1/h)t} + a_2^\star\psi_2^\star e^{2\pi i(W_2/h)t}.$$

To obtain $\Psi\Psi^\star$ we multiply the two together, and obtain one of the most important results of quantum mechanics:

$$\Psi\Psi^\star = a_1a_1^\star\psi_1\psi_1^\star + a_2a_2^\star\psi_2\psi_2^\star + a_1a_2^\star\psi_1\psi_2^\star e^{2\pi i\{(W_2-W_1)/h\}t} + a_1^\star a_2\psi_1^\star\psi_2 e^{-2\pi i\{(W_2-W_1)/h\}t} \qquad \ldots\ldots(36).$$

It will be seen that the frequencies belonging to the separate states have disappeared from the expression, and the pattern possesses instead an oscillation of frequency $(W_2 - W_1)/h$. This is exactly what is needed for incorporating classical and quantum ideas into a new theory. It enables us to use the quantum relation in absorption and emission without renouncing the classical ideas of resonance. However feeble the intensity of incident radiation, an atom cannot absorb less than one quantum. The calculation of an absorption coefficient will be given in Chapter IX.

In the meantime we may enquire what verdict quantum mechanics gives on a question about which the older theories were at variance. Let the arrows in fig. 41 represent two possible quantum transitions from the state W_3 with emission of a spectrum line. It was an essential part of the older quantum theory that at any moment an atom could emit in only one or other of these monochromatic lines. In classical theory, on the other hand, an electron could emit various frequencies at the same time, just as the diaphragm of a gramophone can emit simultaneously tones of different pitch. In the Zeeman effect, for example, where a line has become doubled, in the classical treatment both frequencies are emitted by the same electron; whereas in the Bohr theory it was necessary to insist that one line was being emitted by some of the atoms and the other line by other atoms. To decide this point we must write out $\Psi\Psi^*$ for a state including several levels. In agreement with classical theory many frequencies will be present in the oscillation. But this may be interpreted as meaning that there is a definite *probability* of a quantum of any of these frequencies being emitted. Modern theory intentionally leaves the mechanism vague since it is not accessible to experiment.

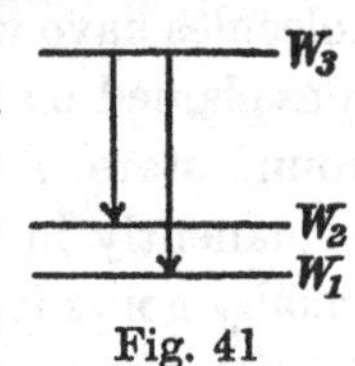

Fig. 41

§5. The original proposal to find a general expression for $\partial\Psi/\partial t$ may now be carried out. Differentiating (33) with respect to the time we obtain

$$\left.\begin{aligned}\frac{\partial\Psi}{\partial t} &= -\frac{2\pi i W}{h}\psi e^{-2\pi i(W/h)t} = -\frac{2\pi i}{h}W\Psi \\ W\Psi &= \frac{ih}{2\pi}\frac{\partial\Psi}{\partial t}\end{aligned}\right\} \quad \text{......(37).}$$

The Schroedinger equation (2) may be written in the form

$$W\Psi - V\Psi + \kappa\frac{\partial^2\Psi}{\partial x^2} = 0 \qquad \text{......(38)},$$

where $\kappa = h^2/8\pi^2 m$. Substituting (37) in (38), we obtain the following pair of equations:

$$\frac{ih}{2\pi}\frac{\partial\Psi}{\partial t} - V\Psi + \kappa\frac{\partial^2\Psi}{\partial x^2} = 0 \qquad \text{......(39)},$$

$$-\frac{ih}{2\pi}\frac{\partial\Psi^*}{\partial t} - V\Psi^* + \kappa\frac{\partial^2\Psi^*}{\partial x^2} = 0 \qquad \text{......(40)}.$$

Equations of this form play the same part in predicting the behaviour of patterns as equations (2) and (9) played in predicting the form of the patterns themselves. The remarkable fact is that (39) contains explicitly $\sqrt{-1}$, which increases the difficulty of attaching a physical reality to the Ψ-waves. But this does not matter, for quantum mechanics recognises that its only object is to predict the probable results of observations. And it becomes increasingly clear that in order to predict the results of atomic experiments it is unnecessary to have any detailed atomic model. To make use of the welcome result of (36), for example, it is not necessary to know what oscillates, nor to suppose that the frequencies $\nu_1, \nu_2, \nu_3, \ldots$ have any particular values, since only differences $(\nu_n - \nu_m)$ are significant. Further, it would have been natural, in introducing the vibrations, to express them by a simple periodic factor, as $\Psi_m = \psi_m \cos 2\pi\nu_m t$; but this is useless because it does not lead to the desired result like (36) in which $(W_m - W_n)/h$ occurs unaccompanied by the frequency $(W_m + W_n)/h$. The use of imaginary quantities is essential, and has the additional advantage that when Ψ is given by (33) the value of $\Psi\Psi^*$ integrated over all space does not fluctuate, but remains equal to unity for all values of t.

§6. It is presumably the uncertainty relation $\Delta W \Delta t \sim h$ which is responsible for the difficulties in giving a clear description of physical events. In all movements of particles we have to be satisfied with a symbolic representation. For this reason the Ψ-waves for free particles have not been considered earlier in this book. We might have expected that the usual form of expression in physics, $A \cos 2\pi (\nu t \pm x/\lambda)$, would represent a travelling plane wave. It will be shown below, however, that we must use the forms

$$\psi = Ae^{\pm 2\pi i x/\lambda}, \quad \Psi = Ae^{-2\pi i (\nu t \pm x/\lambda)} \qquad \text{......(41).}$$

Here the amplitude of the wave A is a kind of normalising factor. For the density of particles in the stream, or the probability of finding a particle in unit length or in unit volume, will be

$$\rho = \Psi\Psi^* = AA^* \qquad \text{......(42).}$$

Consider first a stream of macroscopic particles flowing parallel to the x-axis. If the number of particles crossing every section is the same, i.e. if the current j is the same everywhere, there will be no accumulation of particles anywhere. But if somewhere j varies with x, the density there must be either increasing or decreasing. Consider a small range δx, fig. 42; if the current flowing in is j, that flowing out will be $\left(j + \frac{\partial j}{\partial x}\delta x\right)$. In an interval of time δt a quantity $\frac{\partial j}{\partial x}\delta x\, \delta t$ accumulates in the region, or else there is an equal depletion, according as the value of $\partial j/\partial x$ is negative or positive. In either case the density in the region changes at a rate given by

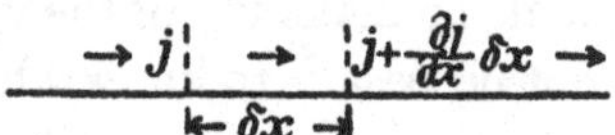

$\rightarrow j \quad \rightarrow \quad j + \frac{\partial j}{\partial x}\delta x \rightarrow$

$\leftarrow \delta x \rightarrow$

Fig. 42

$$\frac{\partial \rho}{\partial t} = -\frac{\partial j}{\partial x} \qquad \text{......(43),}$$

which merely gives expression to the conservation of matter.

To apply this to Ψ-waves we need an expression for $\partial\rho/\partial t$. This is easily obtained from equations (39) and (40). Multiply every term in (39) by Ψ^*, and every term in (40) by Ψ, and subtract. We find

$$\Psi^* \frac{\partial \Psi}{\partial t} + \Psi \frac{\partial \Psi^*}{\partial t} - \frac{ih}{4\pi m}\left(\Psi^* \frac{\partial^2 \Psi}{\partial x^2} - \Psi \frac{\partial^2 \Psi^*}{\partial x^2}\right) = 0 \quad \text{...(44),}$$

which may be written in the form

$$\frac{\partial}{\partial t}\Psi\Psi^* = -\frac{\partial}{\partial x}\left\{\frac{h}{4\pi m i}\left(\Psi^* \frac{\partial \Psi}{\partial x} - \Psi \frac{\partial \Psi^*}{\partial x}\right)\right\} \quad \text{......(45).}$$

Comparing (45) with (43), since $\Psi\Psi^*$ corresponds to ρ, we see that the expression in the large bracket ought to correspond to j. To illustrate this we may take first the simplest motion in field-free space, for which there will be a constant de Broglie wave-length. If (41) represents a uniform stream of particles, the j derived from it ought to be simply the product of ρ and v, the density and velocity of the stream. Taking first the negative sign in (41), and differentiating with respect to x, we find

$$\frac{h}{4\pi mi}\left(\Psi^*\frac{\partial\Psi}{\partial x}-\Psi\frac{\partial\Psi^*}{\partial x}\right)=\frac{h}{4\pi mi}\left(\frac{2\pi iAA^*}{\lambda}+\frac{2\pi iA^*A}{\lambda}\right) \quad \text{...(46)},$$

$$=\frac{A^2h}{m\lambda}$$

$$=\rho v$$

since $\lambda = h/mv$. The current is thus found to have the proper form. Similarly, (41) taken with the other sign will be found to represent a uniform flow in the opposite direction. And when ψ is of the form

$$Ae^{-2\pi ix/\lambda}+Be^{2\pi ix/\lambda} \quad \text{......(47)},$$

where A and B are real, it is easily verified that (46) gives the net flow. In the special case of $B=A$ there will be no net current; in fact, ψ becomes simply $2A\cos 2\pi x/\lambda$, with no imaginary part, so that there is no flow. For owing to the $\sqrt{-1}$ in (45) we can only obtain a real current j when the bracket in (44) and (46) is imaginary.

In dealing with various potential boxes in previous chapters it has been pointed out that in each case a horizontal line lying wholly above the V-curve represents the energy W of a free particle. To such an energy there belongs a Ψ-pattern which is characteristic of the V-curve, and is to be obtained in the usual way by inserting the particular form of V into the Schroedinger equation. There is no quantisation, all values of W giving acceptable patterns, which are oscillatory everywhere. These patterns are important in special problems. For example, when an electron is ejected from an atom, its original pattern becomes replaced by a Ψ-pattern representing the liberated electron moving away.

The other problems in which free motion is important are those of the collisions between particles. These may be divided into two types, according as we are interested in the fate of both particles, or of one particle only. The latter type is naturally the easier to deal with, and embraces such problems as the elastic scattering of a beam of electrons by atoms; we wish to know the angular distribution of the scattered electrons. Starting with a homogeneous stream, (41) will represent a plane wave so long as the value of $(W-V)$ remains constant. If, however, an atom or molecule is introduced into the path of such a wave, the value of $(W-V)$ changes violently in a small region. As usual, Ψ and its derivative must be made continuous at the boundary, and this gives rise to a spherical wave spreading out in all directions, just as a piece of refractory matter placed in a beam of light scatters a feeble spherical wave. The square of the amplitude of the Ψ-wave spreading out in any direction will represent the probability of the incident particle being scattered in that direction.

In introducing the idea of theoretical patterns in Chapter II, attention was drawn to the important fact that the shape of any pattern does not depend on the number of particles which it represents. The Ψ-waves under discussion here afford an illustration of this point. In discussing the flow of a current, it was natural to start with a Ψ-wave representing a stream containing many similar particles. It is clear, however, that the same uniform wave train (41) must be used for representing a *single* particle moving with velocity v in field-free space. A moving particle represented by (41) never gets any farther, for by (42) the value of $\Psi\Psi^*$ remains the same everywhere. This is because the momentum has been given a definite value in λ, and consequently by the uncertainty relation (26) we have no information where the particle is. In order that we may have some knowledge of the position of the moving particle, a Ψ incorporating various energies must be used, as was done for a bound particle in (30). When for a free particle a continuous range of energies is used, as in fig. 33, the composite Ψ is known as a wave packet. The initial spread of this $\Psi(x, y, z)$, that is, the initial accuracy in our knowledge of the particle's position, will depend on how the particle has been prepared.

In most problems we do not require to know the position of a particle at a particular moment. In the scattering of electrons, for example, we wish to know what will happen to an electron which is incident with velocity v; and this incident particle is represented by a plane monochromatic wave train. In the older mechanics one used to say, If an electron is aimed to pass at a distance d from the centre of an atom, it will be scattered through a definite angle θ. Now, however, we have to recognise that we cannot know the initial values of both v and d accurately; hence the wave front of the monochromatic incident wave which we use includes all values of d. By these methods the intensity of scattering from atoms of different elements can be calculated by introducing the appropriate forms of $V(x, y, z)$.

§7. Having seen how to represent moving particles, we are led to re-examine the hydrogen atom with regard to angular momentum. We wish to know whether an electron circulating round the nucleus produces a permanent circular current j. For this purpose we may introduce a complex amplitude into (12), as was done for (2), by writing $B = \pm iA$, and so obtain

$$\Phi = \frac{1}{\sqrt{2\pi}} e^{\pm im\phi} \qquad \text{......(48).}$$

If s is any distance measured along the circumference of a circle of radius r, we have $s/r = \phi$. We obtain a convenient Ψ-wave to introduce into (45) if we consider an electron constrained to move on a definite circle of radius r, writing

$$\Psi(s) = \frac{1}{\sqrt{2\pi r}} e^{-i(2\pi\nu t \pm ls/r)},$$

where l is the quantum number.

We can now use (45) to find the magnitude of the electric current and of its magnetic moment, which is the product of the current and the area of the circuit. Writing μ for the mass of the electron, and taking the positive sign first, we differentiate along the circumference:

$$\frac{h}{4\pi\mu i}\left(\Psi^\star\frac{\partial\Psi}{\partial s}-\Psi\frac{\partial\Psi^\star}{\partial s}\right)=\frac{h}{4\pi\mu i}\left(\frac{il}{2\pi r^2}+\frac{il}{2\pi r^2}\right)$$

$$=l\frac{h}{4\pi^2r^2\mu} \quad \text{......(49).}$$

We thus obtain a circular flow whose value is consistent with an angular momentum equal to $lh/2\pi$. We obtain the current in electromagnetic units on multiplying by ϵ/c. It is this current which, as mentioned in Chapter III, performs a kind of Stern-Gerlach experiment on the electronic spin and causes the familiar splitting of the electronic levels into doublets, triplets, etc. For s-states $l=0$, there is no circular current, and no splitting of the electronic level.

The value of the magnetic moment is obtained on multiplying this current by the area πr^2, with the result

$$l\frac{h}{4\pi^2r^2\mu}\frac{\epsilon\pi r^2}{c}=l\frac{\epsilon h}{4\pi\mu c}.$$

But $\epsilon h/4\pi\mu c$ is exactly the value of one Bohr magneton (the same as the spin moment of the electron itself). The magnetic moment is then either zero (when $l=0$) or an integral number of Bohr magnetons. The same is true of the projection of the magnetic moment on the axis, given by (12). For a Ψ-pattern with which is associated an angular momentum $lh/2\pi$ the resolved magnetic moment may take any integral value from l to $-l$ magnetons. The plus and minus signs in (48), like those in (41), represent electrons circulating in opposite directions. As mentioned in Chapter IV, when an electron with quantum number m and another with quantum number $-m$ are present in an atomic level, their magnetic moments cancel each other out, so that any complete shell is diamagnetic.

§ 8. For a particle bound in a potential box the current j must be zero for each quantised level except in the special case of circulation round an axis, as above. The Ψ-curves for particles merely moving to and fro in a potential box, as in figs. 15, 27, etc., represent standing waves for each allowed level. Nevertheless, it was shown in § 3 of this chapter that when a composite Ψ is formed, the component Ψ's get out of phase with one another, causing the pattern to shift. If this represents the probability of a genuine movement of the particle, one would think it ought to possess meanwhile a real current j, in spite of the fact that for each of the component Ψ-functions, taken separately, j is zero. To test this we may introduce (34) into (45), taking ψ_1, ψ_2, a_1, a_2 as all real quantities. We find

$$\Psi^\star \frac{\partial \Psi}{\partial x} - \Psi \frac{\partial \Psi^\star}{\partial x}$$

$$= \left(a_1 a_2 \psi_1 \frac{d\psi_2}{dx} - a_2 a_1 \psi_2 \frac{d\psi_1}{dx}\right)\left(e^{2\pi i(\nu_2-\nu_1)t} - e^{-2\pi i(\nu_2-\nu_1)t}\right)$$

$$= i a_1 a_2 \left(\psi_1 \frac{d\psi_2}{dx} - \psi_2 \frac{d\psi_1}{dx}\right) \sin 2\pi (\nu_2 - \nu_1) t \qquad \text{......(50).}$$

On multiplying by $ih/4\pi m$ we have a real current whose direction changes periodically. By combining two or more standing waves in this way one can always represent a particle moving in the desired direction during a certain interval, though at some other time the motion is reversed. By means of such a wave packet one can, for example, represent an electron moving about inside an insulated piece of metal.

In the problem of fig. 13, in which $V = cx$ everywhere, it is impossible, taking a particular value of W, to set up a Ψ analogous to (41) to represent a flow parallel to the x-axis. This is what we ought to expect; for although particles need not be reflected at the point where $V = W$, they must be turned back somewhere; hence the flow from right to left must be equal to the flow from left to right, and there are only standing waves. From these a wave packet may be formed, which will move up to the boundary and suffer total reflection.

If, on the other hand, the V-curve AB, instead of rising indefinitely, turns down again somewhere, as BC in fig. 43, then for values of W like that indicated we have a potential barrier, where $W < V$, separating two allowed regions. The same situation occurs in fig. 43b; in this case we know that in the forbidden region ψ is given by (4). Now expressions (46) and (49) suggest that if a complex amplitude were introduced into (4) we should have a steady current flowing through the forbidden region. Writing

$$\Psi = (Ce^{-kx} + iDe^{kx})\, e^{-2\pi i\nu t},$$

with C and D real, and substituting in (45), we obtain as the expression for j

$$\frac{kh}{2\pi m} CD \qquad \text{......(51)},$$

where k is given by (5).

Suppose then that a ψ-wave is incident from the left on to a fairly wide potential barrier (and nothing incident on to the right-hand side). The amplitude of the wave reflected from the barrier will no longer be quite equal to that of the incident wave, giving a small net current to the right, which must also be the value of the current leaking through and emerging on the other side. In Chapter III it was shown that the probability of a particle making an excursion to a distance x beyond the classical boundary was proportional to e^{-2kx}. It is reasonable then to expect that, if d is the width of the potential barrier in fig. 43b, the ratio of the emergent to the incident intensity will be of the order e^{-2kd}. The whole ψ-pattern must be obtained by joining together the various pieces, as was done in figs. 25 to 29, making ψ and $d\psi/dx$ continuous everywhere by adjusting the phases and the values of the constants A, B, C, D, etc. When this is done, it is found that for a given V-curve the emergent current bears a definite ratio to the incident current; except when the barrier is extremely small, this ratio is of the order e^{-2kd}.

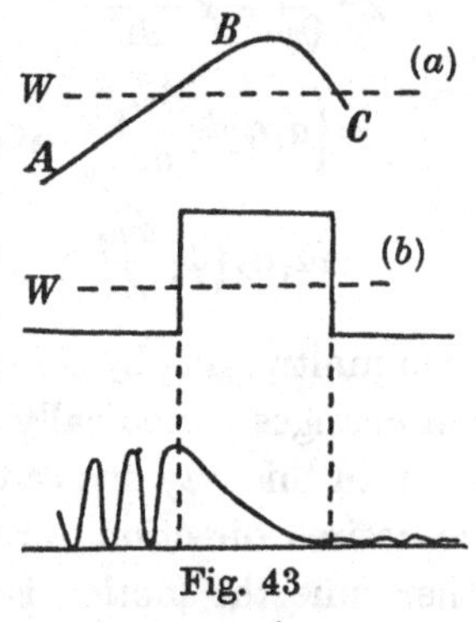

Fig. 43

§ 9. This type of problem will be dealt with in Chapter X. Here it seems better to conclude the chapter by bringing forward a physical process in which a current such as (51) is important. Let us consider the behaviour of the free electrons, when two pieces of metal are brought near together. Since each piece of metal is a potential box, the pair provide a pair of boxes with a potential barrier between, as in the problem of fig. 27, which has been further discussed in § 3 of the present chapter. Electrons making excursions into the forbidden region may from time to time stray from one piece of metal into the other. If the metals are identical the number going in one direction is equal to the number going in the other, and nothing observable happens. If, on the other hand, the two pieces of metal are of different elements, such as copper and zinc, possessing different characteristic work functions, the initial situation will be like that of fig. 44*a*, where the shaded parts represent the occupied electron levels. It is clear that the higher occupied levels in *A* come opposite vacant levels in *B*; so that any leaking of electrons through the gap from *A* to *B* will not be compensated by an equal leak in the opposite direction. The metal *B* therefore acquires a negative charge, and *A* a positive charge, slowly or quickly according to the size of the gap. The growth of these charges means of course the growth of a potential difference; the *V*-curve in the gap acquires a slope which depresses the potential box *A* with respect to *B*, as in fig. 44*b*. The leak from *A* to *B* must continue until the potential difference is sufficient to stop it; which will occur when the electrons in *A* and *B* have been exactly levelled up, fig. 44*b*; the whole will then be in equilibrium.

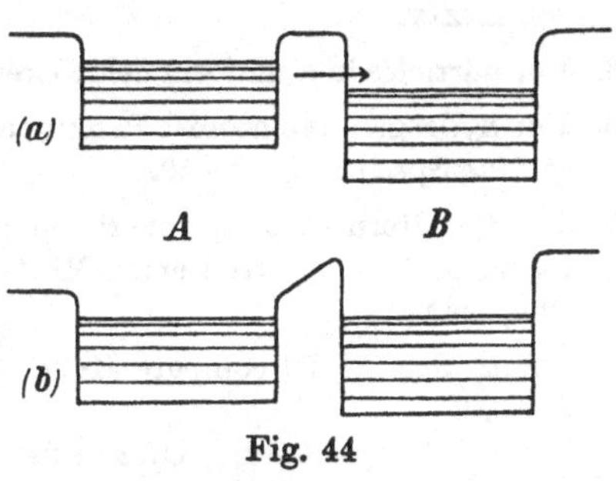

Fig. 44

We have here the origin of the well-known Volta contact potential difference, which arises when any two metals are put in contact. Volta found that the metals could be arranged in a series such that each member became negatively charged with respect to the preceding member in the series. His series is found to agree with a list of the metals in the order of their work functions, in so far as they have been determined by thermionic and photoelectric methods, the value of the contact potential being the difference between the work functions $(\phi_2 - \phi_1)$.

REFERENCES TO CHAPTERS I–V

For general principles see Chapters I and II of Dirac's *Quantum Mechanics*. A mathematical treatment of atomic problems will be found in the *Handbuch der Physik*, vol. XXIV, part I, especially Chapter III.

CHAPTER III

§ 1. Schroedinger's original papers are in the *Annalen d. Physik*, vol. LXXIX.

§ 3. For particles in a uniform field: Breit, *Phys. Rev.* vol. XXXII, p. 273.

§ 6. For hydrogen-like atoms: Pauling and Goudschmidt, *The Structure of Line Spectra*, pp. 25–50.

§ 9. For the Stern-Gerlach experiment: Fraser, *Molecular Rays*, Chapter V; and for electron spin: Mott and Massey, *Atomic Collisions*, Chapter IV.

§ 10. With figs. 27–30 compare Hund, *Zeit. f. Physik*, vol. XL, p. 749, fig. 6.

CHAPTER IV

§ 1. For the Compton effect: Andrade, *The Structure of the Atom*, 3rd edition, p. 688.

For the Heisenberg Principle of Uncertainty: *Zeit. f. Physik*, vol. XLIII, p. 172.

§ 3. For the Pauli Exclusion Principle: *Zeit. f. Physik*, vol. XXXI, p. 765.

§ 4. For the method of the self-consistent field: Hartree, *Proc. Camb. Phil. Soc.* vol. XXIV, p. 120; and Hartree and Black, *Proc. Roy. Soc.* vol. CXXXIX, p. 318.

§ 5. For the Sommerfeld Theory of Metals: *Zeit. f. Physik*, vol. XLVII, pp. 9 and 43; and for the Fermi-Dirac statistics on which it is based: *Zeit. f. Physik*, vol. XXXVI, p. 902, and *Proc. Roy. Soc.* vol. CXII, p. 670.

CHAPTER V

§ 3. With this problem compare Hund, *Zeit. f. Physik*, vol. XLIII, p. 808.

§ 6. For the behaviour of wave packets: Darwin, *Proc. Roy. Soc.* vol. CXVII, p. 258; and vol. CXXIV, p. 375.

For the scattering of electrons: Mott and Massey, *Atomic Collisions*, Chapters VII–XI.

§ 8. For transmission through potential barriers: Gurney and Condon, *Phys. Rev.* vol. XXXIII, p. 130; and Fowler and Nordheim, *Proc. Roy. Soc.* vol. CXIX, p. 173.

§ 9. For the metallic contact: Frenkel, *Phys. Rev.* vol. XXXVI, p. 1604.

CHAPTER VI

§1. TWO INTERACTING PARTICLES

The type of energy diagram introduced in Chapter I has so far sufficed for a simple discussion of each of the problems considered. Even when the system was in three dimensions, as in the hydrogen atom at rest, one could still use a simple V-curve by plotting the potential energy against the radius r. Problems involving the motion of two interacting particles, however, cannot be simplified in this way; reduction to one variable is out of the question. Instead of an energy diagram, one needs at least an energy map. The V-curve becomes a V-surface; and the energy W, instead of being represented by a horizontal line, will now be represented by a horizontal surface.

Taking a one-dimensional problem, suppose that we have a proton and an electron free to move along a straight line. Take any point Q on the line as origin, and let A and B be the momentary positions of the proton and electron respectively. Any such configuration can be represented by a point on an x-y diagram; let the distance of the electron from Q be x, and that of the proton be y. In fig. 45 measure off OC equal to QB, and measure off OD equal to QA. Then the point S represents the configuration when the electron is at B and the proton at A. In the same way every point in the x-y plane represents a possible position for the pair of particles; and any motion of the pair of particles is expressed by a motion of the representative point.

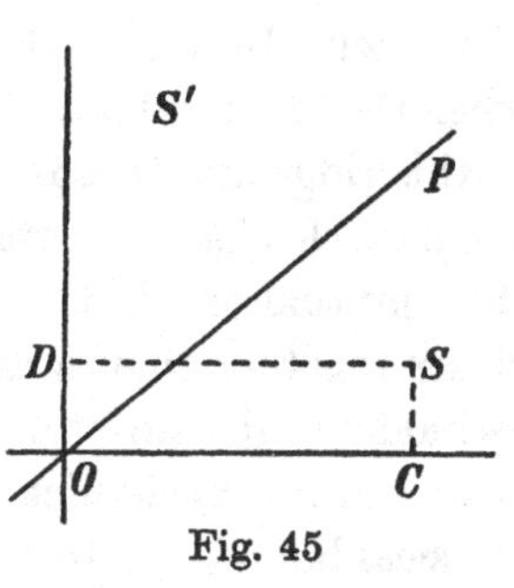

Fig. 45

Along the diagonal OP, drawn at 45° through the origin, we have $x = y$ everywhere. Hence for points near this diagonal the particles are near together, and for points far from the diagonal the particles are far from each other. The V-surface is obtained at once by using the z-co-ordinate to represent the potential energy. When the particles are so far apart that their interaction is negligible, the V-surface is a plane parallel to the plane of the diagram. But all along the diagonal there will be in the V-surface a permanent groove, whose uniform cross-section, taken parallel to the x-axis, is the curve of fig. 7. (If a contour map of the V-surface were drawn, the contour lines would all be straight lines parallel to the diagonal.) The value of W will be represented by a plane parallel to the plane of the diagram. For an electron and a proton bound together the W-plane will cut the V-surface in the groove, and the allowed region will be that portion of the plane lying inside the groove.

If, on the other hand, we start by considering two electrons or two protons making a head-on collision along the line QAB, their mutual energy will be one of repulsion, and the V-surface will be exactly reversed. Along the diagonal OP will be a permanent ridge, whose cross-section has the form of fig. 46. All W-planes cut the V-surface. In classical mechanics, when the two particles collide, the representative point comes up to this ridge and is reflected from it as the particles separate. If the particles have a certain initial relative velocity v, classically their mutual repulsion will prevent them on impact from coming closer together than a definite distance a; under the ridge is a forbidden region, into which the representative point cannot enter. In quantum mechanics there is of course a substitute for this classical boundary. Instead of a representative point we have a Ψ-wave, whose value will die away exponentially into the region under the ridge—meaning that there is a small probability of the particles approaching on impact much closer than the classical distance. The allowed and forbidden regions are evidently not regions of ordinary space, and the Ψ-pattern cannot be pictured as existing in ordinary space.

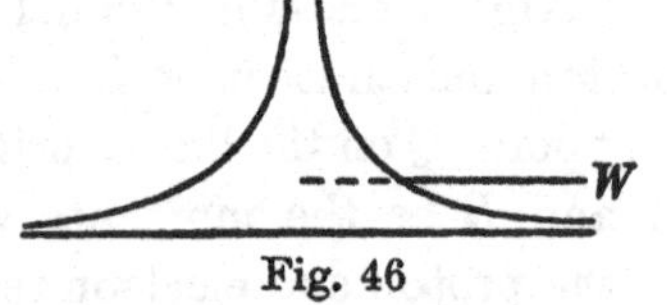

Fig. 46

Hitherto we have been discussing two particles moving in

field-free space. At the other extreme we may consider the case where the potential energy in the external field is very important, for example, when it provides a potential box for the particles. It is convenient to forget about the mutual interaction of the particles for the moment, and to reintroduce it later. Suppose that the potential energy of one particle along the line QRS, fig. 47, is given by curve b. If the charge carried by the other particle is different, it will have a different V-curve. But if the two particles are identical, and both move along the line QRS, the potential energy of both will be given by curve b. Measure off OM and OM' equal to QR, also ON and ON' equal to QS. We obtain a plane V-surface with two similar channels at right angles, one parallel to the x-axis, the other parallel to the y-axis, represented in fig. 47 by the shaded areas. At the intersection of the two channels there is a square box, twice as deep as the original box of curve b. If a section were taken through the V-surface along a line such as EF, it would (in the absence of the diagonal ridge) look like curve d. For any W-plane cutting the lower half of curve d it is only in the square $ABCD$ that $W > V$; when the system has such a value of W, both particles must be bound in the box. (For higher values of W we may have one particle free and the other bound in the box. In what follows we shall consider only the lowest values of W.) Just as in Chapter III we fitted a curve $\psi(x)$ into the potential box of a V-curve given by $V(x)$, so here we shall have to fit a ψ-surface into the potential box $ABCD$ in the V-surface given by $V(x, y)$. To represent this ψ-surface we may again use the z-co-ordinate. Outside the walls

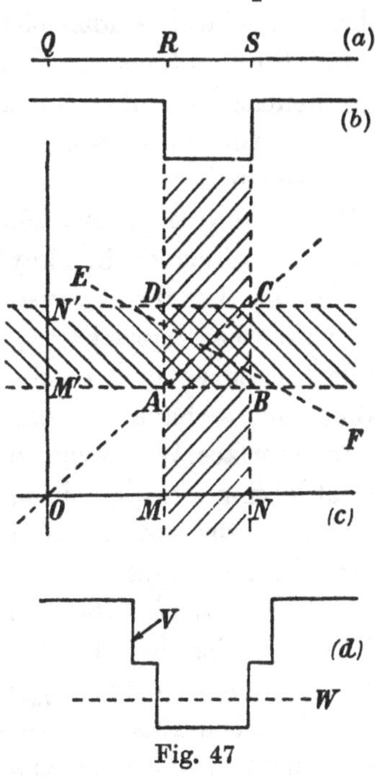

Fig. 47

of the box the value of ψ must fall in all directions exponentially to zero at infinity. There are only certain discrete energies which give ψ-surfaces without kinks. The ψ-surface for the lowest allowed level will be dome-shaped in the box $ABCD$, so that any section through the surface will look like curve a of fig. 15; when this oscillates with its frequency ν, it will alternate between this form and the cup-shaped form whose section is fig. 40b; and so on for the higher levels.

On squaring, ψ^2 gives a probability pattern of less simple type than that discussed in Chapter II. In fig. 8, to represent the probability that the observed quantity lay between q and $q+dq$, we used the vertical strip under the curve; here, on the other hand, we have a volume under a surface. Whenever we are observing two quantities, q_1 and q_2, the volume of a vertical parallelepiped erected on any area $\delta q_1 . \delta q_2$ (fig. 48) gives the probability that the value of q_1 lies in the range δq_1 and the simultaneous value of q_2 in the range δq_2. For the normalisation of ψ it is in this case the whole volume under the ψ^2 surface that must be made equal to unity, for each of the two quantities must certainly have some value. In the problem of fig. 45 the two quantities are the distances of the two particles from the origin, and each particle must certainly be somewhere.

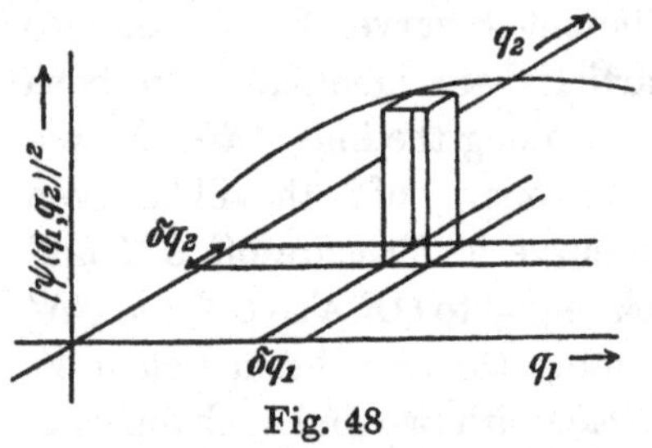

Fig. 48

Reintroducing a repulsion between the two particles (electrons), we shall have again a permanent ridge running along the diagonal of fig. 47 in the V-surface. The effect of this ridge will either be to swamp the potential box $ABCD$, or else to divide it into two equal compartments. If the box is swamped, it means that the mutual repulsion prevents the existence of a stable system. For example the potential box provided by the core of an alkali atom can accommodate one but not two electrons. When, however, the box $ABCD$ is large enough the effect of the ridge is to divide it into two exactly equal compartments, one in *ABC*, the other in *ADC*. Into this pair of compartments we can fit a ψ-surface to represent the two electrons. In this way we obtain a one-dimensional model of the helium atom,* with two electrons confined in the potential box (and of the hydrogen atomic negative ion). The important feature is that the two compartments with the diagonal ridge between introduce the same properties as the boxes *PQ* and *RS* of fig. 27. The ψ-patterns will be all either symmetrical about this diagonal, or else anti-symmetrical.

When the problem has not been simplified by imagining the two electrons restricted to motion along a line, both V and ψ are functions of six co-ordinates. We need the probability of finding one particle in the little volume dv_1 at the point x_1, y_1, z_1, and the other particle simultaneously in the volume dv_2 at the point x_2, y_2, z_2. The ψ-pattern will be in six-dimensional space, and the required probability will be given by $|\psi|^2 dv_1 dv_2$. The symmetrical properties of V and ψ persist in essentially the same form.

An approximate method of dealing with a potential box containing n electrons was mentioned in Chapter IV. We fixed attention on one electron, and imagined the charge of the other $(N-1)$ electrons to be smoothed out, thus determining the average shape of the potential box in which the one electron moved. In this way one obtains an approximate ψ-pattern for each electron separately. On the other hand, the method just described for the helium atom is clearly the correct one, in which we take into account exactly the effect of one electron at x_1, y_1, z_1 upon the other electron at x_2, y_2, z_2. In the lithium atom the three electrons would have to be described by a single ψ-pattern in nine dimensions. The proper treatment of the diffraction of electrons from a crystal surface—the effect which was largely responsible for the development of wave mechanics—is still more complicated. The importance of the method, however, arises from its application to the two electrons in the H_2 molecule, and thence to general problems of valency bonds.

* For orthohelium and parhelium see Mott's *Wave Mechanics*, Chapter VI.

§ 2. DIATOMIC MOLECULES

Atoms that combine to form a molecule have usually been regarded as held together by forces of attraction. One imagined an atom *A* and an atom *B* brought together, and asked whether the forces between them would be attractive or everywhere repulsive. Quantum mechanics now approaches the subject from a different point of view. We start with an atom *A* in a definite quantised state, and an atom *B* in a quantised state, so that the energy of the system is definite. We now suppose that the atoms, initially at rest, are brought nearer together, and we ask, Is the potential energy of the system higher or lower than the initial energy? If it is higher we must have done work in bringing the atoms together; we deduce the fact that the atoms are repelling one another. On the other hand, if the potential energy is lower than the initial value, we detect the presence of an attractive force. (This fixing of attention on the energy is similar to the standpoint adopted in discussing magnetic moments in Chapter III.) On bringing two atoms still closer together, the attraction always changes over somewhere to a more intense repulsion, so that the potential energy rises again steeply, as in fig. 49, where the potential energy is plotted against the distance apart of the atoms. Thus we obtain a potential box, permitting the existence of a stable molecule, the depth of the box being the work required to dissociate the molecule.

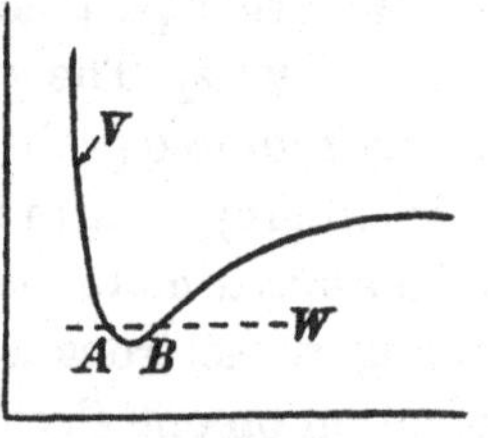

Fig. 49

It will be noticed that here we have come upon a type of problem different from those which have already been treated in this book. In each problem hitherto our *V*-curve was given in advance. Here, on the other hand, we have to use the methods of quantum mechanics to discover whether a stable system is possible, and, if so, what is the shape of the potential box. Empirical *V*-curves of the type of fig. 49 had been in use for some time before the introduction of quantum mechanics, the constants of the curves having been deduced from molecular spectra. There will be one curve for the molecule when the electrons are in their normal lowest state, and a different V-curve for each state where the valence electrons are in an excited level. There may be a marked difference between the curves, as shown in fig. 63, p. 136, since an electron may have quite a different bonding action when it is in a level of higher energy. In this chapter we shall be concerned almost exclusively with the most important V-curve belonging to the normal lowest electronic state. The interpretation of fig. 49 according to classical mechanics would have been as follows: Drawing any horizontal line to represent W, we have inside the box a kinetic energy $(W-V)$. The value of the distance d between the nuclei oscillates continually between two extreme values, such as A and B, depending upon W; for this particular value of the vibration energy, values of d less than A or greater than B are classically forbidden. In quantum mechanics there will of course be a substitute for this classical boundary. But this is anticipating a discussion whose proper place is much later. The first step is to find the potential box.

§3. The simplest molecular system is the ionised hydrogen molecule, H_2^+, consisting of two protons and one electron. We will consider the formation of this molecular ion, starting with a hydrogen atom in its ground state and a proton at a large distance away. The initial ψ-pattern for the electron must be almost exactly that given by (20) for the 1s-level of the H atom, since a sufficiently distant proton can only modify it slightly. Consider, however, what will be the potential energy along the line joining the protons; it will be like curve a of fig. 51. The problem is that of two identical potential boxes, which has already been solved in one dimension in figs. 27–29. The results may be taken over directly into three dimensions. The essential property of the potential energy is that it is symmetrical about the centre of the system. Taking as origin the mid-point between the protons, the value of V at the point $(-x, -y, -z)$ is the same as the value at the point (x, y, z). Hence the value of ψ^2 at any point in one half must be the same as at the corresponding point in the other. But for this it is not necessary that the values of ψ itself should be identical, for they may have opposite signs:

$$\psi(-x, -y, -z) = -\psi(x, y, z).$$

In this way we obtain a set of levels with anti-symmetrical patterns, of which we are only interested in the lowest. For this level the value of ψ taken along any line parallel to the axis would look like curve *c* of fig. 27. To obtain the forms of the patterns and the values of the allowed energies, one would have to insert the proper $V(x, y, z)$ and solve the Schroedinger equation—a problem much more complicated than that of the H atom which occupied the greater part of Chapter III. But the reader will have noticed that curve *b* of fig. 27 is almost the same as would be obtained by adding together the ordinates of the ψ-curves of each potential box taken separately, while curve *c* of fig. 27 is almost the same as would be obtained by subtracting the ordinates. It will be clear from fig. 50 that the value of ψ at Q is going to be rather greater than the value at P in the symmetrical pattern, and

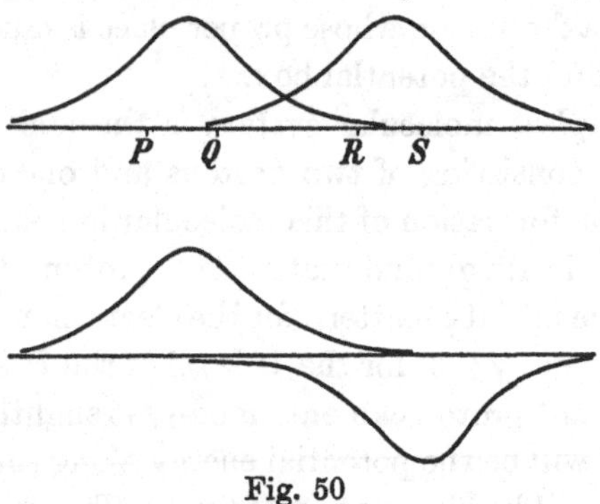

Fig. 50

rather less than at P in the other pattern; so the resemblance is close. If $\psi_A(x)$ denotes the pattern belonging to the box PQ alone, and $\psi_B(x)$ denotes the same pattern belonging to the box RS alone, the distance between the boxes, when both are present, being QR, we have approximately

$$\psi_1 = \frac{1}{\sqrt{2}}(\psi_A + \psi_B) \quad \text{and} \quad \psi_2 = \frac{1}{\sqrt{2}}(\psi_A - \psi_B) \text{......(52)},$$

where the $\sqrt{2}$ has been inserted to preserve the normalisation.

This similarity to the correct patterns is not an accident, but is an example of a general principle, which is discussed in Note 6 of the Appendix, p. 153. It holds for patterns in three dimensions, and will give nearly the correct patterns for the problem under discussion. Take the protons at a definite distance apart, and let ψ_A be the pattern for the $1s$-level of hydrogen with the proton on the left taken as origin, and let ψ_B be the same pattern round the proton on the right. Then, when both protons are present, we have approximately

$$\Psi_1 = \frac{1}{\sqrt{2}} (\psi_A + \psi_B) \, e^{-2\pi i \nu_1 t} \qquad \text{......(53)},$$

$$\Psi_2 = \frac{1}{\sqrt{2}} (\psi_A - \psi_B) \, e^{-2\pi i \nu_2 t} \qquad \text{......(54)}.$$

So long as the protons are far apart the level to which (54) belongs is only slightly higher than the twin level of (53). In either type of pattern the electron cloud is distributed exactly half round one proton and half round the other. This must be reconciled with the fact that, if we start with a hydrogen atom and a proton, the electron is certainly attached initially to one proton or the other. A composite Ψ like that of fig. 35 will evidently meet the case, if, inserting the correct normalising factor, we write

$$\Psi = \frac{1}{\sqrt{2}} (\Psi_1 + \Psi_2),$$

and enquire what will be the form of $\Psi\Psi^*$ at time $t = 0$. From (53) and (54), since ψ_A and ψ_B are both real, we find at time t

$$\Psi\Psi^* = \tfrac{1}{2} [\psi_A{}^2 + \psi_B{}^2 + (\psi_A{}^2 - \psi_B{}^2) \cos 2\pi (\nu_2 - \nu_1) t],$$

which is just $\psi_A{}^2$ at time $t = 0$, representing the electron attached in the $1s$-level to the proton on the left. When the protons are far apart the value of $(\nu_2 - \nu_1)$ is small because the separation of the levels $(W_2 - W_1)$ is small, and the value of the cosine falls extremely slowly from unity. In the same way, by taking $(\Psi_1 - \Psi_2)$ one can represent the electron initially attached to the proton on the right.

As the protons are brought nearer together, we are no longer interested in the momentary position of the electron, but we wish to know accurately the energy as the separation d is made smaller. In the first place the protons repel one another, and work has to be done equal to their Coulomb potential ϵ^2/d, which is plotted in the upper dotted line of fig. 51*c*. At the same time the separation of the twin electron levels increases, as in fig. 31, until it reaches a value of several electron-volts. The method of computing these energies will be given in Chapter IX. When the protons are at any distance d apart, the system will be in one of two alternative states, whose total energies are obtained by adding ϵ^2/d to one or other of the alternative electronic energies.

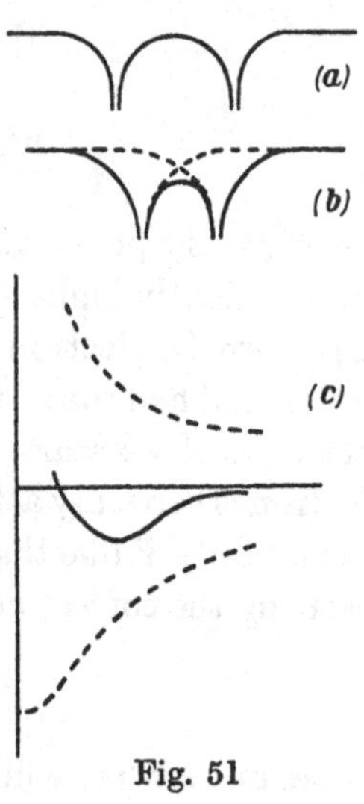

Fig. 51

When the latter have been calculated, one may take in hand the programme proposed in the preceding section, by asking whether over a certain range of d the total energy is less than when the protons were far apart. For the upper of the two alternative states this will clearly not be so anywhere, for the total energy is everywhere increasing. For the lower state it will depend on whether the electronic energy falls more rapidly or less rapidly than the energy of repulsion rises. Calculation for the H_2^+ ion gives the lower dotted curve of fig. 51*c* as the energy of the lower electronic level. Adding together the ordinates of the two dotted curves, one obtains as the total energy the middle curve, which is a potential box permitting the existence of a stable ionised hydrogen molecule. The most important part of the box lies between $d = 0{\cdot}8$ Å. and $1{\cdot}4$ Å.

To sum up, it is the symmetrical electronic ψ-pattern which leads to the stable ground state of the H_2^+ ion. In this single ψ-pattern the electron cloud is distributed half round one proton and half round the other. Although we started with a hydrogen atom and a proton, the electron now belongs to the two protons jointly. Before considering the properties of the H_2^+ ion, we will first examine how to obtain the corresponding potential box for the neutral hydrogen molecule.

§ 4. In the neutral H_2 molecule the two protons provide a pair of identical potential boxes for each electron, exactly as in the H_2^+ positive ion, fig. 51*b*. But since there are now two electrons repelling one another, the considerations of § 1 of this chapter come into force. In order to avoid having to use six dimensions, we first suppose that the electrons move along the line joining the two protons; for this we may use an x-y diagram. Take for simplicity the rectangular potential boxes PQ and RS of fig. 52, initially at a considerable distance apart. With any point A as origin, let the distance of one electron from A be x, and the distance of the other electron from A be y. When we lay off the potential energy parallel to the x-axis, each of the two potential boxes produces in the V-surface an infinitely long channel, running perpendicular to the x-axis. When we lay off parallel to the y-axis the same potential energy for the other electron, each of the boxes produces a permanent channel perpendicular to the y-axis. The mutual repulsion of the electrons produces the usual diagonal ridge OP. At E, where the channels intersect, we have a potential box of twice the original depth, and another at F. If we take a section through the V-surface along any line such as EF, the V-curve so obtained has the symmetrical property of fig. 27. When the protons are held far apart the allowed levels will occur in very close pairs, the ψ-patterns being symmetrical and anti-symmetrical about the diagonal. That is to say, when the ψ-surface is dome-shaped in the box E, it will either be identical in the box F or else cup-shaped with exactly the reverse form.

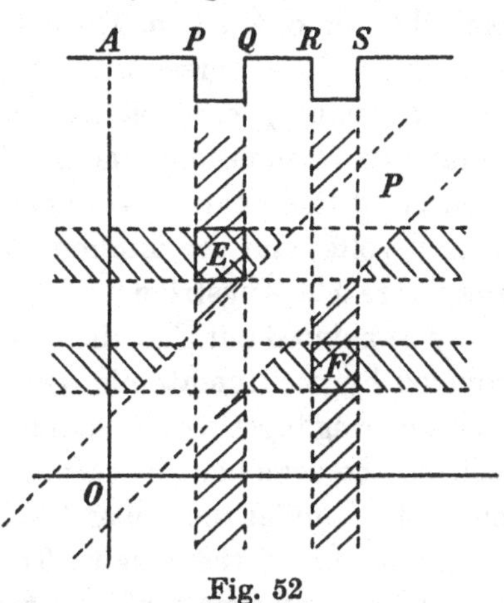

Fig. 52

The rest of the argument runs as for the molecular ion. We bring the protons nearer together, and notice that the splitting of the levels increases exponentially. At the same time there is the Coulomb energy of repulsion to be added. That the upper of the twin levels should lead to a stable molecule is again out of the question; the two hydrogen atoms in this state repel one another at all distances. For the lower of the two levels the result depends on whether the electronic energy decreases more rapidly than the energy of repulsion rises, the total energy of the molecule being the sum of the two. The calculation, which must be made in six dimensions, agrees with observation in providing a curve with a minimum, i.e. a potential box permitting the existence of the familiar H_2 molecule. The depth of the box—the dissociation potential of the molecule—is known from experiment to be 4·4 electron-volts; and the minimum of the curve occurs when the protons are 0·76 Ångström unit apart. In concluding this section, it may be pointed out that the potential boxes of fig. 51*b* may be provided by positive atomic cores instead of by protons. Thus when we bring together two sodium atoms, each with one valence electron, the argument of this section applies. In this way we obtain the familiar molecules Na_2, Li_2, etc., which are common in the vapour of the alkalis. To calculate the shapes of the potential boxes for molecules containing more than two electrons is at present impossible, but they may be deduced empirically from observed band spectra.

§5. Having obtained the potential boxes for H_2 and H_2^+, the next step is to consider the use to be made of them. In contrast to the classical interpretation given in §2, we shall not in any diatomic molecule claim to know the distance apart of the nuclei at any moment with precision. Instead of this, we shall fit ψ-patterns into the potential box, and use ψ^2 to give the probable separation of the nuclei. These ψ-patterns must not be confused with the electronic ψ-patterns with which we have hitherto been dealing.

To find these new ψ-curves, one cannot simply say, Let x be the nuclear separation, and then use equation (2) to find $\psi(x)$. For equation (2) contains m, the mass of the moving particle; and here we are not dealing with one particle, but with the simultaneous motion of both nuclei. This question is discussed in the next chapter, where it is shown that for two particles of the same mass equation (2) may be used as it stands, if we introduce for m half the mass of either particle. The de Broglie wave-lengths which we have to fit into the potential box are by (1) much shorter than those of electrons. Looking for acceptable ψ-patterns for the V-curve of fig. 49, we obtain as a by-product a set of discrete quantised levels, as we did with electrons. The ψ-curve for the lowest allowed energy W_1 will have no node, the next will have one node, and so on as in fig. 15. The lowest levels are nearly equally spaced, as in (8); but for higher levels the width of the potential box increases rapidly and becomes infinite, so that the spacing of the levels becomes narrower and the series tends to a limit.

The difference in energy between the lowest and the next higher vibrational level is very important in connection with the specific heat of the gas. Since the last century it has been a problem why the specific heat of diatomic gases is not larger than the observed values. For the total internal energy per gram-mol at temperature T a value $\frac{7}{2}kT$ was expected, where k is the gas constant, namely $\frac{3}{2}kT$ as translational energy, kT as vibrational and kT as rotational energy, leading to the value $\frac{7}{2}k$ for the specific heat at constant volume. But the experimental value is only $\frac{5}{2}k$. The reason for this becomes clear when we fit the vibrational ψ-patterns into the potential box, and estimate the spacing between the lowest level and the next higher level. The difference in energy turns out to be several times larger than the value of kT, the magnitude of the thermal energy, at room temperature. Consequently in a gas at room temperature, or even above, nearly all the molecules are still in the lowest vibrational level, just as the electrons are in the lowest electronic level; hence the vibrational energy does not appear in the specific heat until quite high temperatures are reached.

§6. In the treatment of the H_2^+ ion in §3 we saw the importance of the splitting of the allowed electron levels when two potential boxes are brought together. If the distance between the potential boxes PQ and RS of fig. 27 is diminished, the V-curve will take the form of curve b of fig. 53, and finally, when the intervening barrier QR disappears, the boxes will coalesce to give the single box PS of curve a. To the V-curve in all these stages there will belong a set of allowed levels and a set of ψ-patterns which will change their shape continuously as the V-curve is altered. At the stage shown by curve b of fig. 53, the patterns belonging to the lowest levels will have the shapes shown in curves c and d. When the boxes were far apart the level to which curve c belongs was the upper of the twin levels into which the ground level split. But now in fig. 53 curve c evidently shows a much greater resemblance to the usual ψ-pattern of the first excited state with one node, of the type shown in fig. 15b, p. 22. In fact, when the boxes coalesce to give a single box PS, curve c will actually become the ψ-pattern of the first excited level of the box, while curve d passes over into the pattern for the ground level. We have thus obtained a quantitative measure of the divergence of the levels. The lower of the twin levels has gone continuously down to become the ground level of the box PS, while the upper has become the next higher level. In the same way, it is easy to see that the curves of fig. 30 will pass over into the patterns belonging to the third and fourth levels.

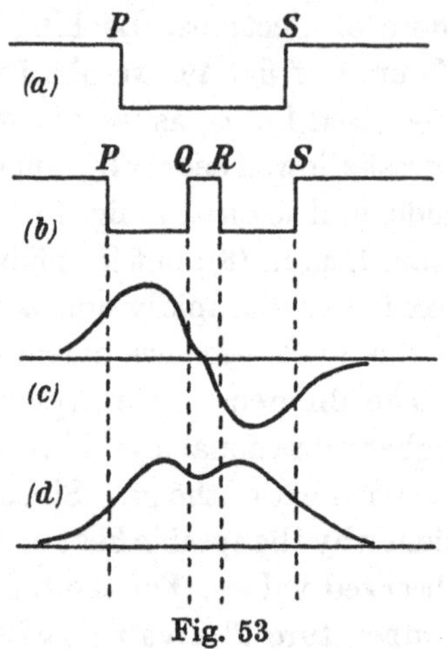

Fig. 53

This suggests a new method of considering the electron levels of molecules and molecular ions. For example, in an H_2^+ ion each of the protons, as we have seen, provides a potential box for the electron. We cannot in practice bring these protons together; but if we could, their potential boxes would coalesce, and the electron would be in the nuclear field of a positive charge 2ϵ; that is, we should have obtained an He^+ ion, all of whose allowed levels we know very accurately, together with their ψ-patterns. As explained in Chapter IV, the ionisation potential of He^+ is Z^2 times the ionisation potential of the hydrogen atom, that is, $4 \times 13 \cdot 5$ ϵ-volts. When the protons are far apart the work required to remove an electron from the system is 13·5 ϵ-volts, the ionisation potential of the hydrogen atom. Hence we know the total fall of the lower dotted curve in fig. 51*c*; it is $3 \times 13 \cdot 5$ or 40·5 ϵ-volts. The shape of the curve can be calculated by the method to be given in Chapter IX, and so the normal unexcited state is determined. If, on the other hand, we start with a proton and a hydrogen atom with its electron in an excited state, we can trace a similar curve connecting it with the corresponding excited level of the He^+ ion. Again adding to this curve the Coulomb energy, as in fig. 51*c*, we should obtain another potential box for the H_2^+ ion, with its electron this time in a particular excited level.

More interesting than the H_2^+ ion is the neutral H_2 molecule, to which a similar argument will apply. Starting with the *V*-surface of fig. 52, we must bring both pairs of channels together, until finally they coalesce, giving a single box at the junction, like *ABCD* of fig. 47. This illustrates the process which must be carried out in six dimensions. If the two protons of a hydrogen molecule could be brought together, we should find the two electrons revolving around a positive charge 2ϵ; that is to say, the electronic ψ-pattern would have become that of a neutral helium atom, whose levels we know. Every level of the H_2 molecule is in a sense intermediate between a level of the H atom and a level of the helium atom. For example, the ionisation potential goes down from 13·5 ϵ-volts for the H atom, through 17 ϵ-volts for the H_2 molecule, to 24·5 ϵ-volts for the He atom. And the systems must have very similar properties, since the electronic ψ-patterns are transformed continuously. In the ground state of the H atom the electrons have no angular momentum; hence they have none in the ground state of H_2 and He. The 1*s*-level of He accommodates two electrons only if they have opposite spin; the spins are antiparallel in the ground state of H_2 also. The electronic ψ-function, not counting spin, for this level of H_2 is symmetrical, as in fig. 27*b*;

it is also symmetrical for the corresponding level of the helium atom. When a molecule is formed from atoms in excited states, the electrons will retain any angular momentum they had initially.

This resemblance of a molecule to the atom which contains the same number of valence electrons is important because it applies to more complicated molecules containing protons, such as the HCl molecule. Let us compare the chlorine atom, of atomic number 17, with the argon atom, of atomic number 18. The chlorine atom has a nuclear charge 17ϵ and 17 electrons, while the argon atom has a nuclear charge 18ϵ and 18 electrons. The HCl molecule will also have 18 electrons, with nuclear charge $(17+1)$. It is reasonable to suppose that the HCl molecule only differs from the argon atom in that one positive charge is separated from the other 17 instead of being united into a single nucleus. Instead of studying the union of a neutral H atom with a neutral Cl atom, we should approach the structure of the HCl molecule by the methods of Chapter IV, using the Exclusion Principle for the 18 electrons, as if we were dealing with a single atom. This treatment, if successfully carried out, should give for the nuclei a V-curve like fig. 49, allowing the existence of the HCl molecule.

The molecular axis of such a diatomic molecule, i.e. the line passing through the nuclei, is of course not fixed in space. On the contrary, we shall allow the proton to rotate about the chlorine nucleus, just as in Chapter III we allowed the electron to rotate about the hydrogen nucleus. In fig. 18 the original Coulomb curve became modified by the rotation, to give other potential boxes; so here the original V-curve will become modified by various amounts. In no diatomic molecule does one nucleus really rotate about the other, but both of course rotate about their common centre of gravity. And, strictly speaking, it was incorrect in Chapter III to treat the electron in the H atom as rotating round the proton; for both really rotate about their common centre of gravity. The whole treatment, however, becomes valid if the r and μ denote respectively the distance apart of the particles and the "reduced mass", $m_1 m_2/(m_1+m_2)$, which was already used in classical mechanics and in Bohr's theory.

CHAPTER VII

§ 1. INDEPENDENT OBSERVABLE QUANTITIES

Let the probabilities of the occurrence of various events be denoted by $P_1, P_2, P_3, \ldots$; then if these events are independent of each other, the probability of their occurrence together is given by their product

$$P = P_1 P_2 P_3 \ldots.$$

This at any rate is a theorem which is not undermined by quantum mechanics, and may be applied to the kind of probabilities in which we are interested, namely the kind which give the probabilities that certain observable quantities, say, $q, r, s, \ldots$, have values lying simultaneously in particular ranges $dq, dr, ds, \ldots$. In Chapter II it was pointed out that in any complicated system there are many quantities that can be measured—distances, angles, etc.—and that, when we cannot get definite values, we need patterns to describe all these observables. The above theorem suggests how the state of a complex system may be described, the only difference being that we shall use square roots of P, expressing our probabilities in the usual form $|\psi|^2$. In any special case, then, where the quantities q, r, s are independent of one another, we may take the product

$$|\psi_q|^2 dq \, |\psi_r|^2 dr \, |\psi_s|^2 ds,$$

which is equal to

$$|\psi_q \cdot \psi_r \cdot \psi_s|^2 dq\,dr\,ds \qquad \ldots\ldots(55).$$

Hence this system is to be described by a composite ψ which is the product of three ψ-functions:

$$\psi(q, r, s) = \psi_q \psi_r \psi_s \qquad \ldots\ldots(56).$$

This ψ may be contrasted with the composite ψ which was obtained in Chapter IV by adding together various ψ-functions. The $\psi_1, \psi_2, \ldots$ which were added together were different functions of the same variable, and all had the same dimensions. On the other hand, the $\psi_q \psi_r \psi_s$ which are multiplied together in (56) are functions of different variables q, r, s, and will not in general have the same dimensions. As usual in physics, we add together like quantities and multiply together unlike quantities. It is obvious that, whatever dimensions q, r and s may have, the quantity given by (55) is always a pure number.

The kinetic energy possessed by a complicated system at any moment is obviously the sum of the kinetic energies possessed by its parts; and the same is true of the potential energy. The total energy W must be the sum of the sum of W_q, W_r and W_s. The use of (56) will make nonsense unless multiplying together ψ-functions is compatible with adding together energies. We see at once that the two are made consistent by the property of exponents, since

$$\Psi_q \Psi_r \Psi_s = \psi_q \psi_r \psi_s e^{-2\pi i \{(W_q + W_r + W_s)/h\} t}.$$

In earlier chapters it has not been necessary to mention these aspects of ψ-patterns, because we were fixing attention, as far as possible, on one observable quantity at a time. But in the last chapter, in describing a diatomic molecule, we had found the need of one ψ-pattern, which may be called ψ_{el}, to describe electrons, and another ψ-pattern to express the distance apart of the nuclei, and there is also the rotation of the molecular axis to be dealt with. It will be best to review the whole chapter from the beginning.

Starting in § 1 with two charged particles moving along a straight line, attention was fixed in fig. 45 on their distance from some origin. The probability of finding one particle near a certain point depended upon whether the other particle was at or near that point, or not. Therefore the $\psi(x, y)$ which we were to obtain from fig. 45 could not be derived as a product $\psi_a(x)\psi_b(y)$, as the two quantities were not independent; see Note 4 of the Appendix. If it were possible, it would clearly be to our advantage to pick out quantities to measure which specify the configuration, and yet are independent of each other. In the case of two interacting particles moving in otherwise field-free space this can be achieved by fixing attention on (a) the distance of the centre of gravity of the pair from some origin, and (*b*) the distance apart of the particles from each other. See Note 5 of the Appendix. For example, if in fig. 45 OD had represented the distance of the centre of gravity from the origin, and OC the distance apart, then the point S would have represented a particular configuration of the pair; and all possible configurations can be represented by other points S. For all points near the y-axis, on either side of it, the value of x is small, and so the potential ridge or groove, as the case may be, which in fig. 45 ran down the diagonal in the V-surface, now runs down the y-axis. Motion in the y-direction represents motion of the centre of gravity, while motion in the x-direction represents the two particles approaching or separating. For example, if the centre of gravity is known to be moving along a line with velocity v given by $h/mv=\lambda$ (where m is the sum of the masses), we shall use

$$\Psi=\psi(x)\,.\,Ae^{-2\pi i(\nu t\pm y/\lambda)}.$$

In this way, by making use of (20), a moving hydrogen atom may be represented by

$$\Psi=\frac{A}{(\pi a)^{3/2}}e^{-(2\pi iWt/h+r/a\pm 2\pi iy/\lambda)},$$

where W includes the kinetic energy $\frac{1}{2}mv^2$ of the atom. In this case we have no information as to the momentary position of the atom (i.e. of the centre of gravity); where we have such information $\Psi(y)$ must be a wave packet instead of a monochromatic wave.

Any standing ψ-waves incident on and reflected from the permanent potential ridge in the V-surface will represent two particles colliding; from such standing waves a wave packet moving parallel to the x-axis may also be formed. When the two particles are in otherwise field-free space, the potential ridge has the same cross-section everywhere; in this case there is no need to use a V-surface, for one may obtain $\psi(x)$ and $\psi(y)$ separately from the two corresponding V-curves. The proper V-curve for finding the separation of the particles is clearly one in which the potential energy is plotted against the distance r between them. And this is just the type of V-curve whose derivation for a molecule was considered in figs. 49–51. As already explained, by fitting patterns into such a potential box we obtain a function $\psi(r)$ for the nuclear separation. And at the end of the last chapter we saw the need of a $\psi(\theta, \phi)$ to give the probability of finding the molecular axis lying in any direction θ, ϕ. In the absence of any external field the nuclear separation is quite independent of the orientation of the axis in space. Hence, when we have found ψ_{vib} for the nuclear vibration and ψ_{rot} for the rotation of the molecule, a composite ψ may be obtained by multiplying the two together:

$$\psi = \psi_{\text{vib}} \psi_{\text{rot}}.$$

Although we did not approach the subject from this point of view, we see that it was in agreement with this procedure that for the electron and proton in the hydrogen atom we obtained the pattern as a product

$$R(r) . \Theta(\theta) . \Phi(\phi).$$

Thus $\Theta . \Phi$ gives the orientation of what may be called the atomic axis as it rotates about the centre of gravity. The molecular problem is in fact the same, and must be solved by means of equations (78) to (85). A constant, $C = j(j+1)$, where j is integral, must be introduced, leading to quantisation of the energy of molecular rotation. The initial potential box becomes modified by the subtraction of various amounts of rotational energy $j(j+1)h^2/8\pi^2\mu$. But whereas in the H atom the reduced mass μ was almost the mass of the electron, for the nuclei of a diatomic molecule it is of course much larger, so that the potential boxes for various values of j are almost identical with each other and with the original box for $j=0$. The result is that, when we fit ψ-curves into each of these boxes, the nth level of any box almost coincides with the nth level of any other box. These small energy differences, when we form $\Psi\Psi^*$ for a composite Ψ, as in (36), give rise to oscillations of very low frequency, and are responsible for very long infra-red wave-lengths that are absorbed and emitted when the molecule jumps from one rotational level to another.

In this paragraph we have shown that the behaviour of the two nuclei in a diatomic molecule is quite similar to the behaviour of the two particles which form a hydrogen atom. We have said nothing about the electrons in the molecule. The ψ_{el} for a rotating molecule is the same as for a stationary molecule, which has already been discussed in the last chapter.

Part of the scheme of levels for a vibrating molecule is shown in fig. 54. The levels shown all belong to one particular electronic level of the molecule—say the lowest electronic level. To each of the other electronic levels will belong a similar set of vibration-rotation levels. The range of energies covered by the diagram is intended to be less than one ϵ-volt, i.e. small compared with the energies of electronic excitation. If this difference of scale is borne in mind, fig. 54 may be compared with fig. 20. In either diagram the spots in any vertical row belong to the same angular quantum number, i.e. to a particular modified potential box. The lowest level in any vertical row has no node, the next has one, which becomes a spherical nodal surface, and so on; a change of vibration involves a change in the number of these, while a rotational transition leaves their number unchanged. (It might have been asked why, if the problems are similar, the electron in a hydrogen atom does not emit vibrational and rotational spectrum lines. The answer is that it does; only from fig. 20 it is clearly not profitable to distinguish between the two types, because they are mixed up together.)

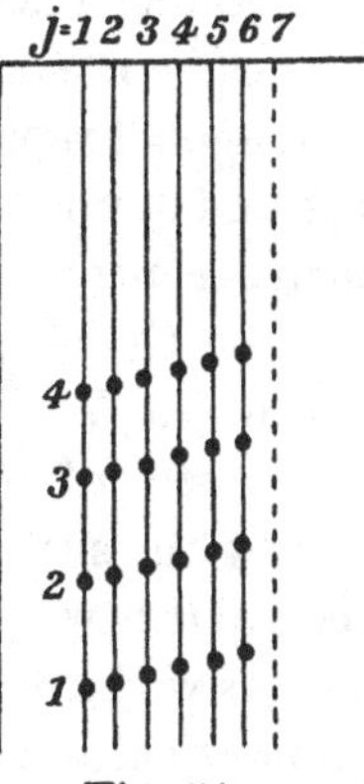

Fig. 54

§ 2. HOMONUCLEAR MOLECULES

Diatomic molecules whose nuclei are identical are called homonuclear. Although in fig. 52 identical potential boxes were drawn, in order to represent H_2, the identity of the boxes did not play any part in the discussion. For consider what the V-surface would be like if the boxes were of different width. Laying off the potential energy parallel to the x-axis and parallel to the y-axis, we obtain two pairs of channels in each of which one channel is wider than the other. The potential boxes at E and F now become rectangular in plan, instead of square. But the important point is that they lie symmetrically with respect to the diagonal; in fact the entire V-surface is by construction completely symmetrical about the diagonal. Hence again $|\psi_{el}|^2$ must be symmetrical but ψ_{el} may be anti-symmetrical. Whereas in the H_2^+ ion these symmetry properties depended on the identity of the protons, here they depend on the identity of the two valence electrons.

Taking two particles which we label 1 and 2, consider fig. 45 once more. The point S represents the situation when, say, particle number 1 is at A on the line QAB, and particle number 2 is at B. Now the corresponding point S' lying on the other side of the diagonal represents the situation when particle number 2 is at A and number 1 is at B. If the particles are identical, the two labels are meaningless, and the probabilities of the two situations must be the same, since they are not distinguishable by any experiment. We may therefore make the following remark about any system of whose particles two are identical: Every ψ describing the system must be such that when we interchange the labels of the two particles ψ^2 remains unaltered; but ψ may change sign. As we have seen in fig. 51, the difference in energy between symmetrical and anti-symmetrical levels is often very important. The forces accompanying these energies were an innovation in quantum mechanics, and the name "exchange forces" has been given to them, because they arise in situations where we suppose identical particles to be exchanged or interchanged.

Since the material universe is supposed to consist mainly of electrons and protons, the H atom is perhaps the only non-nuclear stable system which does *not* contain identical particles. Other ways in which the identity enters into molecular physics must at least be considered briefly.

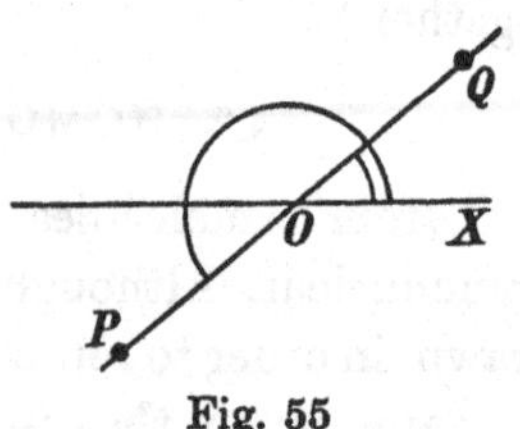

Fig. 55

Take for example a rotating homonuclear molecule, with reference to fig. 17, and consider the projection of the molecular axis. In fig. 55 the two nuclei have been labelled P and Q; let ϕ be the angle which OP makes with OX at any moment; in the diagram ϕ is about 210°. If now we interchange the labels P and Q, the angle XOP will be 30° or 390°, i.e. $(\phi \pm \pi)$. We must now recall that if ϕ is any angle, and m is any odd integer,

$$\cos m(\phi \pm \pi) = -\cos m\phi,$$
$$\sin m(\phi \pm \pi) = -\sin m\phi.$$

If therefore we use (48), namely $\Phi = A \cos m\phi + iA \sin m\phi$, we see that for those levels for which the quantum number m is odd, when we interchange the labels, Φ changes sign, and is therefore anti-symmetrical. On the other hand, those levels for which m is even have a symmetrical Φ, since interchanging the nuclei leaves its sign unchanged. It can also be shown that we have symmetrical and anti-symmetrical patterns according as the rotational quantum number j is even or odd.

This property has a large and unexpected effect on the specific heat of hydrogen. When we cool H_2 down towards absolute zero, we should ordinarily expect nearly all the molecules to fall into the lowest rotational level, i.e. with $m = 0$. But it can be shown that we should expect transitions from any anti-symmetrical state to any symmetrical state to be very infrequent; and observation supports this view in various phenomena. Consequently, although molecules initially in even rotational levels will go into the level with $m = 0$, those initially in odd levels will go first into the lowest anti-symmetrical level, i.e. with $m = 1$, and will take days to change over to the lower level. Thus hydrogen behaves like a mixture of two gases; the two forms have been called orthohydrogen and parahydrogen. It will obviously take more heat to warm up the H_2 from the level $m = 0$ than from $m = 1$, so that the two forms have different specific heats, the usual specific heat being that of the mixture. As this difference in specific heat is accompanied by a difference in vapour pressure, it is actually possible to separate the two forms from each other.

§ 3. Consider now the properties of a composite ψ which is a product of two ψ's each of which has symmetry properties, such as $\psi=\psi_p\psi_q$. Consider a particular level for which both ψ_p and ψ_q are anti-symmetrical. When we interchange the particles ψ_p changes sign, and also ψ_q changes sign. The product $\psi_p\psi_q$ has not changed sign, so that the complete ψ for this level is symmetrical. This may be expressed in the form $A.A=S$. On the other hand, when either ψ_p or ψ_q is anti-symmetrical, but not both, the complete ψ is anti-symmetrical, that is, $A.S=A$ and $S.A=A$. The same argument will apply to a product of three ψ's, the sign following the ordinary rules of algebra.

Consider now the various excited states of a homonuclear molecule which give rise to its band spectrum. We may leave ψ_{vib} out of account because it is always S and so does not affect the result. The ψ_{el} will be S for some electronic levels and A for others, while ψ_{rot} is A or S according as the rotational quantum number is odd or even. We should have expected, then, that in any molecule both kinds of complete ψ would be represented among the levels, namely,

$$\left.\begin{matrix} S.S \\ A.A \end{matrix}\right\}=S, \quad \left.\begin{matrix} S.A \\ A.S \end{matrix}\right\}=A \qquad \text{......(57).}$$

From the observed levels, however, it appears that homonuclear molecules of different elements may be divided into two classes. For some elements only those patterns occur in which the complete ψ is A, while for others only those patterns occur whose complete ψ is S; by a kind of exclusion principle no molecule possesses both kinds. This has important consequences, depending upon whether the nuclei of the molecule possess internal spin momentum.

It was mentioned in Chapter III that the proton possesses a small spin angular momentum. In the H_2 molecule the spins of the two protons will lie sometimes parallel to each other, and sometimes anti-parallel; this has no appreciable effect on the values of the allowed energies. The directions of the nuclear spins in any molecule must be expressed by a factor ψ_{ns}, which may or may not change sign on interchanging the nuclei. The character of the complete ψ is now given by the product of three factors

$$\psi = \psi_{el}\, \psi_{rot}\, \psi_{ns}.$$

Some atomic nuclei, such as He^4, C^{12}, O^{16}, do not possess spin; and for any of these there will be no ψ_{ns} to be added. It is clear now that homonuclear molecules formed from such atoms without nuclear spin, such as common O_2, will, according to the rule given above, possess only half as many levels as a molecule with nuclear spins. Alternate rotational levels will not occur in nature because by (57) they give a complete ψ of the wrong kind. For molecules with internal nuclear spin there is no restriction, because $\psi_{el}\psi_{rot}$ may be combined with any ψ_{ns} which is A or with any ψ_{ns} which is S, whichever is required.

Our knowledge of these molecular levels is of course obtained from the band spectra emitted in quantum transitions. It is observed that for molecules such as $O^{16}O^{16}$ each band is the same as the corresponding band for an ordinary molecule, except that every other line is missing—"alternate missing lines". Molecules which show this effect are found to be those whose atoms from other evidence are believed to possess no nuclear spin; hence the above rule has been deduced.

In parahydrogen all molecules have their nuclear spins anti-parallel, so that these contribute nothing to the magnetic moment. In orthohydrogen two-thirds of the molecules have their nuclear spins parallel, and one-third anti-parallel. By means of a Stern-Gerlach experiment on a stream of orthohydrogen it should thus be possible to determine the value of the magnetic spin moment of the proton. Frisch and Stern in 1933 obtained a value 2·5 times the expected value; i.e. it was about 1/730 of a magneton, instead of 1/1838. A similar experiment with the heavy isotope of hydrogen gave for the magnetic moment of the deuton or diplon a value about one-third of that of the proton.

§ 4. VALENCE BONDS

The whole discussion so far has thrown very little light on the fundamental question as to why some atoms unite to form stable molecules, while others do not. The core of the problem is obviously the existence or non-existence of a V-curve with a minimum, like fig. 49. But is there any guiding principle by which the form of the V-curve may be estimated? And in particular, what is the connection between the new methods and the empirical rules which have been in vogue since the last century, namely, the rules of valency? There must be some connection, because the valency rules were satisfactory, apart from certain important exceptions.

Let us begin by asking why a helium atom and a hydrogen atom will not combine to form a molecule HeH. Why cannot a curve like that of fig. 49 be obtained? The electrons in a helium atom in its ground state are both 1s-electrons, and that in a hydrogen atom is also a 1s-electron. By the Exclusion Principle no system may contain more than two 1s-electrons (with opposite spin). When therefore we try to form an HeH molecule by bringing the atoms together, it is essential that one of the electrons shall be raised towards a higher level, namely a 2-quantum level. We have already seen in fig. 53 how a level may pass over gradually into a level of higher quantum number. When such a level is occupied by an electron, the process is known as the "promotion" of the electron. In the H_2 problem of fig. 52 the descending level was capable of accommodating the electrons of both combining atoms. But here when we bring together a helium atom and a hydrogen atom, there is an electron left over, which must go into an ascending level. As far as the positive nuclei and two of the electrons are concerned, the process is similar to that in H_2, i.e. the repulsion of the former is counteracted by the lowering of the level of the latter. The question is whether the work necessary to promote the third electron is sufficient to turn the scales against the formation of a molecule. The amount by which the valence level is raised may be estimated by an argument like that used in the last chapter. If the proton and the helium nucleus did not repel one another, we could bring them together to form a single nucleus. Starting with a hydrogen and a helium atom, we should finally have three electrons circulating round a positive charge 3ϵ; that is, we should have a lithium atom, whose ionisation potential is only 5 ϵ-volts. As a result of the promotion the valence level is more than 8 ϵ-volts higher than it was in the hydrogen atom. Whether H and He repel one another at all distances can only be decided by detailed calculation. A similar argument applies when we bring together two helium atoms in their ground state. There are four 1*s*-electrons, two of which must be promoted, leading to a repulsion.

It has already been mentioned on p. 94 that the single valence electron in a lithium atom will behave like the electron in a hydrogen atom. On bringing together two lithium atoms the lower of the twin levels into which the valence level splits leads to an attraction between the atoms. We must not, however, neglect the fact that the atoms together contain four 1*s*-electrons, two of which must be promoted, as in the case of helium. We must satisfy ourselves that the consequent repulsion is small compared with the attraction due to the bonding action of the 2*s*-electron; then we shall be justified in calling the 2*s*-electron the valence electron of the atom. Now it was pointed out with regard to fig. 31 that promotion or its converse begins to be appreciable when the distance d between the atoms is small enough to give an appreciable value to e^{-kd}; and we recall that the factor k contains $(V-W)$ the height of the potential barrier between the boxes. The K-level of the lithium atom is of course a low-lying level, about 60 volts lower than that of the 2*s*-electron. Hence the factor e^{-kd} which determines the promotion of the K-electrons is negligible until d becomes very small, and hence does not interfere with the formation of the observed Li_2 molecule, whose normal nuclear separation is as great as 2·6 Ångström units. This reasoning is applicable to low-lying levels in all atoms; we are justified then in regarding the loosely bound electrons in an atom as alone concerned with valency.

But loosely bound electrons often contribute a repulsion between atoms, as we shall see in the molecules N_2, NO and O_2. To deal with these atoms we may use the notation explained in Chapter IV. The configurations from carbon to neon are:

Carbon	$1s^2 . 2s^2 . 2p^2$
Nitrogen	$1s^2 . 2s^2 . 2p^3$
Oxygen	$1s^2 . 2s^2 . 2p^4$
Fluorine	$1s^2 . 2s^2 . 2p^5$
Neon	$1s^2 . 2s^2 . 2p^6$

Neon possesses a closed shell because only eight electrons are allowed in the 2-quantum level. When, however, we bring together two nitrogen atoms, we have initially ten electrons in the 2-quantum level, namely five in each atom. Consequently two of these electrons will clearly need to be promoted towards a 3-quantum level. The fact that a stable molecule is formed is due to the fact that this repulsion is exceeded by the attraction due to the bonding action of the other eight electrons. The formation of the molecule may be described by saying that each atom contributes four bonding electrons and one anti-bonding electron; the latter cancels out one of the bonds, leaving three bonds.

In the formation of the NO molecule there are initially eleven electrons in the 2-quantum level, three of which need to be promoted. And in the formation of O_2 there are twelve, of which four need to be promoted; each oxygen atom contributes four bonding electrons and two anti-bonding electrons, leaving two bonds. We should expect that the presence of these promoted electrons would lead to a smaller value for both the ionisation potential and the energy of dissociation of O_2 and NO than for N_2 and smaller still for fluorine F_2. The observed values show that this is so.

The molecules HF, H_2O, NH_3 and CH_4 with ten electrons each must all have the same electron configuration as the neon atom. The radical CH has the same electron configuration as the nitrogen atom. It is not surprising then that two CH radicals combine to form acetylene C_2H_2 just as two nitrogen atoms combine to form N_2. In the same way the boron hydride radical BH_3 with eight electrons has the same configuration as the oxygen atom, and so has CH_2. Consequently they are found only in the forms B_2H_6 and C_2H_4, corresponding to O_2, while the molecule C_2H_6 corresponds to fluorine F_2.

REFERENCES TO CHAPTERS VI AND VII

Chapter VI

For the experimental approach to molecular states: Smyth, *Rev. Modern Physics*, vol. I, p. 374.

For the Heitler and London method: *Zeit. f. Physik*, vol. XLIV, p. 453.

Chapter VII

For mathematical treatment: *Handbuch der Physik*, vol. XXIV, part I; and Kronig, *Band Spectra and Molecular Structure*.

§ 2. For Ortho- and Para-hydrogen: Dennison, *Proc. Roy. Soc.* vol. CXV, p. 483; and Farkas, *Ergebnisse d. Exakten Naturwissenschaften*, vol. XII, p. 163.

§ 3. For homonuclear molecules: Mulliken, *Rev. Modern Physics*, vol. III, p. 147.

For the magnetic moment of the proton: Frisch and Stern, *Zeit. f. Physik*, vol. LXXXV, p. 4, and vol. LXXXIX, p. 665.

§ 4. For Mulliken's work on valency: *Rev. Modern Physics*, vol. IV, p. 1. And for further work on molecules, see references to Chapter IX.

CHAPTER VIII

§ 1. ELECTRONS IN CRYSTALS

The potential energy of an electron along any line passing through a positive atomic core has the form of fig. 7, while that taken along a line passing through two identical atomic positive cores has the form of fig. 51*b*. Now in any crystal, such as diamond or silver for example, there are from the point of view of a valence electron rows of identical positive cores equally spaced throughout the lattice. The V-curve will therefore be like curve *a* of fig. 56, if the line taken passes through the atomic nuclei, and like curve *b*, if it does not. The essential features of the problem may first be analysed by replacing this V-curve by the simpler curve *c*. To investigate the behaviour to be expected of electrons in crystals, we enquire what will be the forms of their ψ-patterns belonging to various energies W. The problem falls for analysis into two parts: (*a*) where the periodic variation of curve *c* is supposed to be repeated *ad infinitum* in either direction, and (*b*) where an actual piece of crystal possesses definite boundaries. We shall begin with (*a*), and introduce the boundaries of the lattice later.

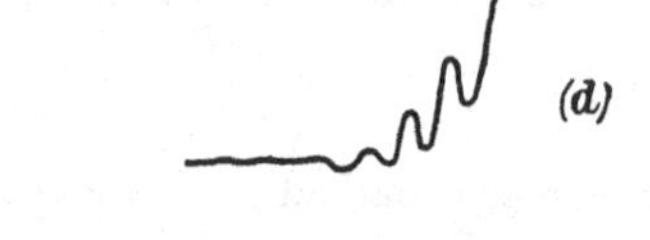

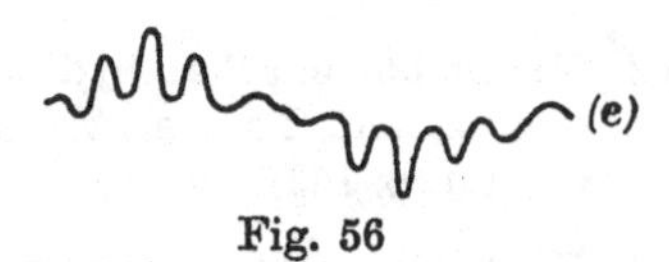

Fig. 56

Taking any curve such as curve *c*, we draw as usual a horizontal line to represent W; and we distinguish between a line which lies wholly above the V-curve, and one which cuts it repeatedly. But this distinction, usually of such fundamental significance, is found in this problem to have lost much of its importance. This fading out of the customary division gives rise to the most interesting features of the problem. (1) Taking first a W-line lying wholly above the V-curve, we shall have possible patterns representing a flow to the right or to the left. The kinetic energy of the moving particle $(W-V)$ will vary periodically, but to any particular value of W there corresponds a definite speed of motion through the lattice. Any ψ-curve for curve c is obtained by fitting together fragments of sine curves of the two different wave-lengths involved; the curves for the real and imaginary parts are obtained separately. (2) Turning next to a W-line which cuts the V-curve repeatedly, we see that there is no reason why the ψ-patterns should necessarily be of a different nature from the former. Although space is divided up into small allowed regions, the width of the intervening barriers is never more than

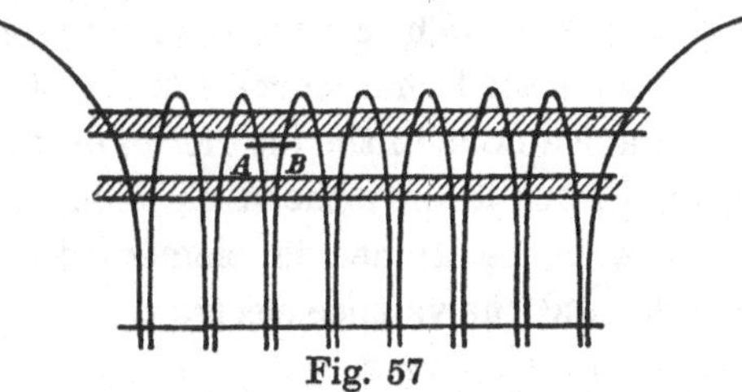

Fig. 57

about 2 or 3 Ångström units (we know this from the lattice spacing of crystals), and such thin barriers present scarcely any obstacle to the motion of electrons, unless the barrier is very high. The ψ-patterns are to be obtained by the method of figs. 28 and 29, adjusting the amplitudes of the fragments of sine curve and exponentials to give a continuous smooth curve. This could always be done if we did not mind a lavish use of an increasing exponential. But of course we are restricted as usual to patterns which remain finite everywhere. For some energies the only possible patterns have an amplitude which goes on increasing indefinitely in one direction (curve d) while decreasing indefinitely in the other; these are to be rejected. In acceptable patterns the amplitude may oscillate to any extent (curve e), provided that it remains finite. An investigation shows that these two types of patterns occur in successive zones of energy, which consequently become alternating allowed and forbidden zones of energy. This is illustrated in fig. 57 if we suppose that the shaded zones represent energies where all patterns remain finite, while in the intermediate zones there are no patterns representing electrons resident in the crystal. Each crystal will possess its own characteristic zones, depending on its lattice spacing and upon the particular form of the field provided by the positive atomic cores. In the infinitely large crystal which we have been considering all energies within the allowed zones are possible. For an electron confined within a finite piece of crystal, on the other hand, the value of ψ must die away exponentially from the boundaries in all directions. As for any potential box this is only possible for certain discrete energies. This quantisation selects from each of the allowed zones a definite number of levels, depending upon the size of the piece of crystal. (The ψ-patterns for valence electrons are really more complicated than curve e of fig. 56. In lithium the pattern must have a node in the interior of every atom, since by the Exclusion Principle only the K-electrons may have no node in the potential boxes. In heavier elements the L-electrons will have one node, and the valence electrons must have two or more in the interior of each atom.)

§ 2. The problem of the levels may be approached from a different angle. Starting with the one atom, we may imagine the row of atoms of fig. 57 to be built up by adding similar atoms one by one (an extension of the process of building a diatomic molecule). We find that each atom contributes its own levels to the crystal, so that when the row contains N atoms there are exactly N levels in each allowed band, each level being able to accommodate two electrons with opposite spin. When all the electrons have been inserted into the levels in these bands, some of the bands will be filled while those above will remain empty. We can see now the connection between this model and the model of Chapter IV. The bottom of the simple potential box of fig. 37 corresponds to the bottom of one of these allowed bands, namely that occupied by the valence electrons, below which is a wide zone of forbidden energies. In lithium metal there is a gap of nearly 60 ϵ-volts between the K-levels and the wide band occupied by the valence electrons. With this potential barrier, nearly 60

ϵ-volts high, between adjacent atoms, the K-electrons are immobile and may be treated as attached to particular atoms, as in the Li_2 molecule. In fact, the main result of this discussion is to show that every crystal is nothing more than a large molecule. In Chapter VI it was emphasised that in a diatomic molecule the valence electrons do not belong to one atom or the other, but circulate around both cores. So here the ψ-patterns for the valence electrons are not localised round particular atoms, but extend uniformly through the whole crystal. A standing wave such as curve e of fig. 56 is not essentially different from the simple de Broglie standing waves which were used for metallic electrons in Chapter IV. In both cases a composite Ψ may be formed from them, representing a current j flowing at any moment through the crystal, as explained in § 8 of Chapter V.

§ 3. INSULATORS AND CONDUCTORS

Very surprising at first sight is the fact that these considerations must apply to insulating crystals as well as to metals. That we are right in supposing a free flow of electrons through an insulator to be possible is shown by the phenomenon of photoconductivity. Many good insulators when illuminated with light of suitable wave-length exhibit electronic conduction, showing that electrons may travel through the lattice. The problem is to explain why they are good insulators in the dark. In classical mechanics there was no difficulty, because no motion through the interatomic regions where $W < V$ was possible; one merely had to say that in an insulator no electrons were free to move. But in quantum mechanics there is a much greater resemblance between an insulator and a metal than had previously been supposed.

In a crystal the electrons are normally streaming uniformly in all directions. When a voltage is applied to a metal, some of the electrons are accelerated in the direction of the field. Any electron which has acquired kinetic energy in this way has jumped to a level of slightly higher energy. Consider a crystal of an alkali metal containing N atoms. According to the Exclusion Principle, the N levels in the band containing the valence electrons can accommodate N pairs of electrons with opposite spin, that is $2N$ electrons. Since each atom possesses one valence electron only, this band of levels will be exactly half full. The situation at zero temperature and at room temperature is similar to fig. 38, since the finite width of the band is not affecting the distribution. Immediately above the occupied levels are plenty of vacant levels into which electrons can jump when accelerated. Hence alkali crystals must be good conductors of electricity at all temperatures. It is important to recognise, however, that only a small fraction of the valence electrons are accelerated by the field and help to carry the current, namely those in the highest occupied levels of the valence band. The energy that can be acquired from the applied field is quite small, and at ordinary low temperatures the electrons in the lower parts of the band have almost no vacant levels to go to. Consequently, the number of conduction electrons is much smaller than the number of valence electrons.

Going on now to the alkaline earths, we see that since each atom possesses two valence electrons, there will be exactly enough electrons in any crystal to fill an allowed band of levels. This at once gives a clue to the existence of insulating crystals (in fact, at first sight one would expect the alkaline earths to be insulators). In all the familiar insulators (*a*) the number of electrons is such as to exactly fill a certain number of allowed bands of levels, and (*b*) the forbidden zone of energy separating these occupied levels from the next higher vacant band is of considerable width. Although the ψ-patterns of the electrons extend through the whole crystal, it is impossible to use them for conduction. An applied field does not accelerate the electrons because there are no levels of slightly higher energy to which the accelerated electrons can jump. From applied fields of ordinary intensity an electron cannot acquire enough energy to jump to the next allowed band. The essential feature of an insulator then is that the valence electrons form a closed group in the crystal.

The metallic nature of the alkaline earths depends on a property of the system of levels which has not yet been mentioned.

It has been pointed out that in a one-dimensional row of atoms there are always forbidden zones separating the allowed bands, the position and width of these zones depending upon the spacing of the atoms. In a three-dimensional crystal the spacing between the lattice planes is different in different directions. In a cubic crystal, for example, the (100) planes are farther apart than the (110) or the (111) planes. In any crystal each complete band comprises levels representing motion in all directions. Hence it may happen that the top of one band of levels overlaps the bottom of the next higher band, i.e. there is no intervening zone in which *all* directions of motion are forbidden. In the range of energy where the bands overlap, the electrons will be distributed partly among one kind of level and partly among the other, and above them there are plenty of vacant levels. Such a crystal will be metallic. In an insulator the allowed bands must be completely separated.

The motion of electrons through a crystal will be unimpeded only when the lattice is perfect. At every temperature except absolute zero the V-curve of fig. 56 is no longer completely regular but is subject to distortion by the thermal vibrations of the lattice. When a current flows through a metal the motion of the electrons is continually being checked, and a mean free path must be assigned to the electrons, depending on the temperature. The energy acquired from the accelerating field is given back to the lattice and appears as heat. The electrical resistance of metals, which arises in this way, is small at low temperatures, and increases rapidly as the temperature is raised. The presence of impurities in a metal has the effect of increasing the resistance, because the regularity of the periodic V-curve is thereby diminished. In an insulator the presence of impurities has the opposite effect, as will be explained later.

Turning next to the optical properties of crystals, we see that the proposed scheme of levels explains why insulators are usually transparent. The transparency and the insulating property arise from the same cause, namely, that electron transitions are only possible from one band of levels to the next. A pure crystal cannot absorb wave-lengths whose $h\nu$ is less than the width of the intervening zone of energy. Transparent crystals always exhibit characteristic absorption in the ultra-violet region, as we should expect. From the wave-lengths at which this absorption occurs we deduce the width of the zone between the allowed levels. For any crystal which is transparent to all visible light the gap must be at least 3 ϵ-volts. For those crystals which are transparent into the Schumann region of the ultra-violet the gap must be at least 5 or 6 ϵ-volts.

§ 4. Atoms present as impurities in such a crystal play an important rôle, even if present in very small quantity. For there is no reason why the electron levels of such an atom should lie within the allowed bands belonging to the lattice; usually they lie in the empty zone of energy between the lattice bands. Hence they can absorb wave-lengths other than those characteristic of the lattice, and give to the crystal a feeble continuous absorption spectrum. Consider the ψ-pattern for an impurity level such as AB of fig. 57 lying within an otherwise empty zone; the amplitude of ψ will be large only inside the impurity atom itself. Hence the level is localised, and an electron occupying the level cannot move through the lattice. Yet the presence of impurities has an important effect in diminishing the insulating power of a crystal. Imperfections in the lattice and sub-microscopic cracks will give rise to similar localised levels even in a pure crystal.

Fig. 38 showed how in a metal at any temperature a certain number of electrons are thrown up into higher levels. Similarly in an insulator the effect of the thermal vibrations will be to keep a few electrons up in the otherwise vacant band of levels above the forbidden zone. These electrons will be accelerated by any applied voltage, but their motion will not give any detectable current at room temperature, because their numbers are so small; calculation by the Boltzmann relation shows that when the width of the forbidden zone is 2 ϵ-volts, it is unlikely that there will be even one such running electron in any cubic centimetre. This may be contrasted with the behaviour of the valence electron of an impurity atom whose valence level happens to lie just below this nearly empty band as AB in fig. 57; by the Boltzmann relation the probability of the electron being jerked up into the conduction levels of this nearly empty band may be many million times greater. Hence even at room temperature a very small number of such impurity atoms will supply enough conduction electrons to carry a measurable current. The higher the temperature the greater will be the number of such electrons kept in the running levels. And it is a well-known fact that the resistance of insulators diminishes with increasing temperature, in contrast to that of metals. It is now thought that the whole class of solids intermediate in properties between metals and insulators, and known as semi-conductors, would be good insulators if they could be obtained in a pure form. Ordinarily their conduction electrons are supplied by impurity atoms whose valence levels lie above, in, or just below, the empty conduction levels. And it is established that the resistance of such substances is higher the purer the specimen.

In view of the band of levels through which electrons can run freely, there is still a question to be answered. Why, when an insulating crystal is placed between two pieces of metal, cannot the free electrons of one metal be caused to run through the insulator into the other metal? To answer this question properly one must consider what happens when an insulating crystal is placed in contact with a metallic crystal, before any voltage is applied. The situation is analogous to fig. 44. Electrons from the metal cannot make transitions to the full band of the insulator nor to the forbidden zone of energies, so we have only to fix attention on the metallic levels opposite to the nearly empty running levels of the insulator. As the two crystals are first brought together, the number of electrons in these levels of the metal will be either greater or less than the few electrons present in the nearly empty band. If it is less, electrons will make transitions to the metal, charging the latter negatively, and raising the metallic levels, until there is equilibrium. If it is greater, a few electrons will make transitions to the crystal, charging its surface negatively, thereby lowering the levels of the metal, until the number of electrons opposite the running band is as small as in the band itself. When a voltage is applied these electrons may stream through the insulator, but they are not more plentiful than the initial running electrons of the insulator. Most of the conduction electrons of the metal are a long way below the empty running levels of the insulator, and there is no mechanism for raising them.

§ 5. Returning to the row of atoms of fig. 57, consider the potential energy of the last atom at the right-hand end. If we push the atom to the left, there is an intense repulsion. If we pull the atom to the right, there is an attraction which falls off as the distance increases. The V-curve for this atom is in fact like fig. 49. The work required to detach an atom from a crystal surface is the sublimation energy, corresponding to the dissociation energy of a molecule. In the case of a metallic crystal we may detach instead a positive atomic core, leaving behind N electrons and $N-1$ cores; for this there will be another V-curve like fig. 49. There will be a set of vibration levels for the V-curve, as for a molecule.

Still fixing attention on the last atom at the right-hand end of the row, let us ask to what extent the situation would be different if this one atom belonged to a different element, i.e. if we had a foreign atom adsorbed on to the surface of a metal. If the original crystal is just a super-molecule, the final state is merely a super-molecule containing another atom. The V-curve for the electrons will still be like fig. 57, except that (a) the mean distance of the adsorbed atom from its neighbour may be different, and (b) the internal field provided by its atomic core may be different. When we determine the shapes of the electronic ψ-patterns belonging to this V-curve, the amplitude which ψ has at this end of the row may be greater than before for some energies, and less than before for other energies. When we have a monatomic layer of foreign atoms adsorbed on to a metal surface, the total density of the electron cloud may be less than normal in the adsorbed layer, or greater than normal. When it is less, we shall find at the surface an electrical double layer, positive outwards; when it is greater, we shall have a double layer, negative outwards.

When an electric charge is conveyed through a double layer, work is done or else energy is liberated: i.e. there is an additional step up or down in the V-curve, as in curves a and c of fig. 58. For an electron at a metal surface on which a layer is adsorbed, curves a and c may be combined with the V-curve of fig. 37. Thus we obtain curve b when the layer is negative outwards; the value of the work function for electrons is effectively increased. We should expect the position of the photo-electric threshold to be affected by the presence of adsorbed atoms. Both the photo-electric and thermionic effects are in fact extremely sensitive to contamination of the surface. It is found by experiment that oxygen adsorbed on to a surface of tungsten metal raises the work function from 4·4 ϵ-volts to more than 8 ϵ-volts. On the other hand, when the double layer is positive outwards, the work function is reduced. In curve d we have a narrow potential barrier through which the Ψ-wave can leak. It is in this way that the filaments of dull-emitted valves emit a large thermionic current at a fairly low temperature.

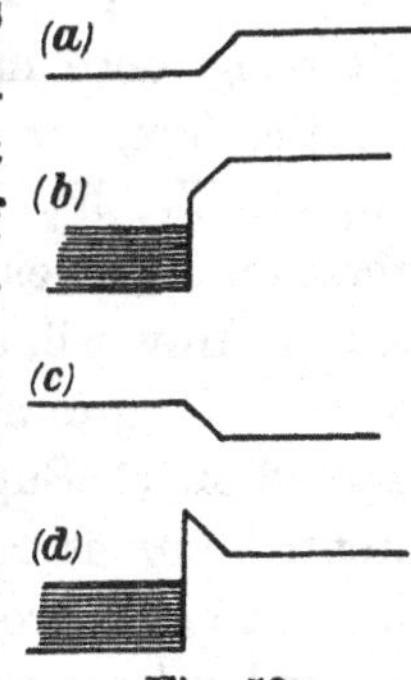

Fig. 58

REFERENCES TO CHAPTER VIII

§1. For the Bloch treatment of motion through a lattice: *Zeit. f. Physik*, vol. LII, p. 555.

Kronig and Penney, *Proc. Roy. Soc.* vol. CXXX, p. 499.

§2. Brillouin, *Quantenstatistik*.

Peierls, *Ergebnisse d. Exakten Naturw.* vol. XI, p. 265.

O'Bryan and Skinner, *Phys. Rev.* vol. XLV, p. 374; Jones, Mott, and Skinner, *ibid.* vol. XLV, p. 379; Jones, *Proc. Roy. Soc.* vol. CXLIV, p. 225.

For the optical properties of metals: Mott and Zener, *Proc. Camb. Phil. Soc.* vol. XXX, p. 249.

§4. For semi-conductors: Wilson, *Proc. Roy. Soc.* vol. CXXXIV, p. 279.

For photo-conductivity: Robertson, Fox, and Martin, *Phil. Trans.* vol. CCXXXII, p. 507.

§5. Hughes and Dubridge, *Photoelectric Phenomena*.

CHAPTER IX

§ 1. PERTURBATION THEORY

Starting once more with a charged particle moving in a potential box, for which we know the values of the allowed energies, W_1, W_2, W_3, ..., suppose we make some permanent alteration in the electrostatic field in a region near or including the box. How will the values of the allowed energies be altered? To clear the ground, there are two simple cases which can be disposed of. (1) Suppose that the whole V-curve containing the potential box is raised or lowered by a uniform amount $\mathcal{V}$, relative to our zero of energy. The allowed values of W will all be raised or lowered by an equal amount $\mathcal{V}$, since the acceptable ψ-patterns depend only on the value of $(W - V)$. (2) At the other extreme suppose that the alterations of the V-curve were made in a region throughout which the value of one or more of the ψ's were zero. The allowed energies belonging to these ψ's would be unaffected. Strictly speaking, neither of these cases can be exactly realised, since every ψ-pattern occupies the whole of space. But we can say what comes to the same thing, that if the alteration in V is made in a region where the value of any ψ is negligibly small, the modification of the allowed energy to which this pattern belongs will be negligible. We can lay it down then as a first obvious principle (*a*) that in the following discussion only regions where the value of ψ is appreciable will be taken into account. And, returning to the first problem, we can say in case (1) that if the uniform alteration $\mathcal{V}$ embraces all but a negligible fraction of any pattern, the change in the allowed energy will be almost exactly equal to $\mathcal{V}$.

The supposition, made above, that the alteration $\mathcal{V}$ should be uniform throughout a certain region is of course highly artificial. It may be used for illustrating the principles, just as we have often used a potential box of artificial shape, but in practice the alteration will always vary from point to point. To the original potential energy $V(x, y, z)$ is added $\mathcal{V}(x, y, z)$, so that the resultant potential energy is $(V+\mathcal{V})$. Often the effect of the alteration may be to raise the V-curve in some places and to depress it in others. On the other hand, the alteration *may* be everywhere of the same sign, varying in magnitude from point to point. At any rate we may say that if in the regions given by (*a*) as relevant the alteration in V is everywhere positive, the shift of every level W_n will be positive; and if in the relevant regions $\mathcal{V}$ is everywhere negative, the shift of every W_n will be negative. This agrees with what we already know of allowed energies. We know, for example, that any change which makes a potential box narrower will raise all the levels, while any change which makes the box wider will lower them.

Having cleared the ground in this way, the next step will be more easily made by considering an actual example. Suppose that the alteration in V is due to the arrival of a free proton which was initially absent. The alteration, being $-\epsilon^2/r$, is everywhere negative. It extends through the whole of space, but is negligibly small outside a certain spherical volume. We may lay it down then as a second obvious principle (*b*) that only regions where the value of $\mathcal{V}$ is appreciable will be taken into consideration in determining the shift of the allowed energies W_n. Combining this idea with (*a*) we are led to what may be called the Principle of Overlapping, which plays many different rôles in quantum mechanics, as explained below. Every region where ψ_n and $\mathcal{V}$ appreciably overlap contributes to the shift in the corresponding W_n. The contribution made by an element of volume $dx\,dy\,dz$ depends upon the value which $\mathcal{V}$ has, and which the normalised $|\psi_n|^2$ has in that volume. The shift in the allowed energy is obtained by integrating over all space. If W_n' is the new value of W_n, the shift is to a first approximation

$$W_n' - W_n = \int \mathcal{V} |\psi_n|^2 dv = \int \psi_n \mathcal{V} \psi_n^* dv \qquad \text{......(58).}$$

The derivation of (58) is given in Note 7 of the Appendix. It will be noticed that in the case of complete uniform overlapping this leads by (11) to the result of case (1) given above.

A simple example of incomplete overlapping is given in fig. 59 for the sake of illustration. Let curve *a* be the initial *V*-curve, and let curve *c* be the pattern belonging to the lowest level of the box. Now let the potential energy be modified by the application of curve *b*; the alteration is to be zero except in the range *PQ*, where it is equal to $\mathcal{V}$; the effect is just to deepen the potential box by an amount $\mathcal{V}$. All the levels will be lowered, but since the alteration overlaps only the central portion of each pattern, the lowering of each level should be proportionately less than $\mathcal{V}$ if (58) is correct. Since the modified potential box has higher walls, the discussion of fig. 23 on p. 42 is sufficient to verify the conclusion without further comment.

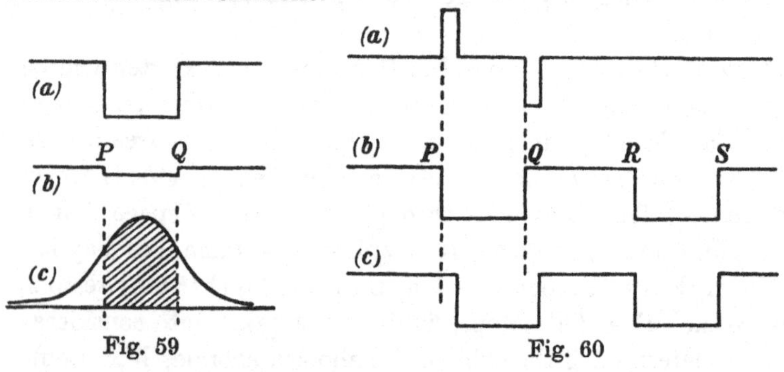

Fig. 59

Fig. 60

The usefulness of (58) becomes evident when we remember how quantum mechanics regards the forces between atoms. To find the magnitude of the repulsion or attraction between two atoms, we suppose the distance between them altered, and enquire how much the energy has been raised or lowered. Equation (58) is clearly an instrument for this purpose. As an illustration we may apply it to the simplified H_2^+ ion, using a pair of rectangular boxes. Since we already know the result for this simple case by direct consideration of the patterns, we shall be able to test the use of (58). Starting then with curve *b* of fig. 60 as our original *V*-curve, we wish to push the potential boxes nearer together. The necessary alteration in the potential energy is evidently of the form of curve *a*. On adding the ordinates of curves *a* and *b*, the effect is to fill up a portion of the potential box at *P*, and to carve out an equal portion from the potential barrier QR, with the result that the box PQ is shifted sideways to the right, i.e. nearer to the box RS, as shown in curve c. Curve a is then our $\mathfrak{V}(x)$, the alteration being zero everywhere except near P and Q. The integral (58) will have to be taken only over these two small ranges, and the only portions of the ψ-curve which will enter into the calculation will be the values near P and near Q.

The negative value of $\mathfrak{V}$ near Q is of course equal to its positive value near P, both being equal to the depth of the potential box. The value of the integral (58) would therefore be zero if ψ^2 had exactly the same values near Q as near P. It has, however, been sufficiently emphasised with reference to figs. 27 and 50, that in every anti-symmetrical pattern the value of ψ is slightly smaller at Q than at P, while in every symmetrical pattern it is slightly smaller at P than at Q. The familiar splitting of the levels follows at once, for the integral (58) will be positive for each anti-symmetrical pattern, raising the allowed level, and negative for each symmetrical pattern, lowering the allowed level as in fig. 31. If the values of any ψ at P and at Q are denoted by $\psi(P)$ and $\psi(Q)$, then if the distance between the boxes is reduced by an amount dx, we have

$$\frac{dW}{dx} = \mathfrak{V}\{[\psi(P)]^2 - [\psi(Q)]^2\}.$$

It is easily verified that the right-hand side of this equation has the dimensions of a force, i.e. mass times acceleration. This example gives some idea of the calculation which must be made in three dimensions to obtain by this method the shape of the dotted curve in fig. 51c for the H_2^+ ion.

§2. The method for the neutral H_2 molecule may be illustrated in the same way. The V-surface of fig. 52 must clearly be modified by means of a $\mathfrak{V}$-surface in such a way as to bring each pair of channels closer together. The integral (58) must then be taken over the relevant strips of the x-y surface. We need to know the approximate electronic ψ-surface to be introduced into (58). In contrast to fig. 50, we cannot of course obtain the approximate ψ-pattern for the molecule by adding together the separate patterns for the electron in each atom. Knowing the electronic ψ-pattern for each of the separate atoms, we must decide on the proper method of combining them. In the ψ-surface of fig. 52 we have seen that the value of ψ must be small except in the potential boxes E and F, which are the two regions where $W > V$. As in Chapter VI, let ψ_A be the pattern for the box PQ, and ψ_B that for the box RS when there is a definite distance QR between the boxes. Then let us form the product $\psi_A(x)\psi_B(y)$. This product has a definite value at every point of the x-y diagram; and we see that its value is large only in the region E and falls away exponentially outside the box owing to the exponential decrease of one factor or the other, or of both. In the same way it is clear that the product $\psi_B(x)\psi_A(y)$ is large only in the region F and is small everywhere else. Suppose then that we take the ψ-surface

$$\psi_1 = \psi_A(x)\psi_B(y) + \psi_B(x)\psi_A(y) \qquad \text{......(59).}$$

This surface will be dome-shaped in the boxes E and F, falling to a small value between them and elsewhere. In the same way the ψ-surface

$$\psi_2 = \psi_A(x)\psi_B(y) - \psi_B(x)\psi_A(y) \qquad \text{......(60)}$$

will be dome-shaped in the box E and cup-shaped in F, with exactly the reverse form. But these are respectively the characters which the symmetrical and anti-symmetrical patterns of lowest energy for fig. 52 must have. In fact (59) and (60) represent the first step in combining the electronic patterns of the separate atoms, when the atoms are far apart. As we have entirely neglected the presence of the diagonal ridge in the V-surface, it is at first sight a little surprising that (59) and (60) resemble the correct patterns when the ridge is present. But this depends upon the fact that when the atoms are far apart, the value of ψ is very small near the diagonal, and provides an illustration of the Principle of Overlapping, i.e. that alterations in the potential energy are unimportant if they occur only in those regions where ψ is very small.

The above illustrates the method which in the genuine problem of two atoms must be carried out using six co-ordinates, x_1, y_1, z_1, x_2, y_2, z_2. For two atoms whose ground states are s-states we know what we mean by ψ_A and ψ_B as soon as we have chosen a particular distance apart of the nuclei. For comparatively few elements, however, is the ground state an s-state. For the others the electronic ψ-pattern of the ground state is not spherically symmetrical. For any two patterns taken from fig. 10 the combined pattern will clearly depend on their relative orientation. Fig. 61 represents an atom in an s-state and an atom in a p-state at a fixed distance apart but differently oriented. If the two atoms combine to form a molecule when brought together, their stable configuration will be that which provides the V-curve with the lowest minimum; the same is true of three or more atoms. Although the necessary calculations are laborious, this treatment opens up a very important field—the spatial orientation of valency bonds and the actual shapes of polyatomic molecules.

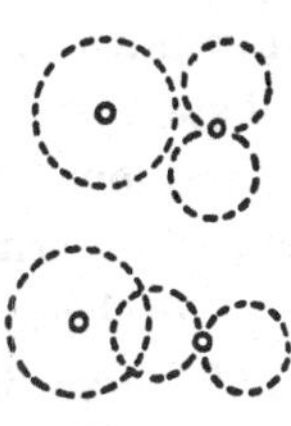

Fig. 61

§ 3. At the beginning of this chapter we supposed that we knew the allowed energies and the ψ-patterns for a particular V-curve, and we set out to enquire whether, using the material at our disposal, we could find the allowed energies W_n' of a slightly different V-curve. An extension of this problem is obviously to enquire whether we can, again using the material at our disposal, find also the new patterns ψ_n'. Consider, for example, the pattern belonging to the level W_2'; this will differ slightly from the original ψ_2. And the idea is that we could modify ψ_2 in the right direction, by adding to it small quantities of the other original ψ-curves which we already know, writing

$$\psi_2' = \psi_2 + a_1\psi_1 + a_3\psi_3 + \dots \qquad \text{......(61)},$$

where a_1, a_3, ... are small numbers, positive or negative. The problem is to know the best possible values of these numbers to insert. The rule for calculating them is derived in Note 7 of the Appendix.

§ 4. THE DESCRIPTION OF PHYSICAL EVENTS

At the beginning of Chapter v the excitation of an atom was cited as an example of a physical process. For a given disturbance there will be no certainty to which excited state the atom will be raised, but only a set of probabilities. If the atom is known to be initially in the ground state, the value of a_1 will be unity until time $t = 0$. Then the Ψ-patterns of the excited levels will start growing at the expense of the pattern belonging to the ground level. This is merely a particular illustration of a general type of description. The growth of Ψ-patterns at the expense of the initial pattern or patterns, when some alteration is made in the V-curve, is the standard representation of a physical event in quantum mechanics. As in § 1, $\mathfrak{V}$ is added to the initial potential energy V, to give $(V + \mathfrak{V})$.

When we ask, Which Ψ-patterns will grow most rapidly? we can begin to clear the ground, as in § 1, by a few obvious preliminary answers. In the first place, if the alteration $\mathfrak{V}$ is made in regions where the value of the initial Ψ_1 is negligibly small, nothing measurable will happen at all. Secondly, if there were any Ψ's which did not overlap the initial Ψ_1 at all, these need not be taken into account, because there would be no possibility of the particle represented by the initial Ψ_1 making a transition to the other state. Since, however, every pattern occupies the whole of space, this rule must be put in the form: Any Ψ which does not anywhere overlap the initial Ψ_1 appreciably need not be taken into account. On the other hand, the patterns which have the largest overlapping with the initial pattern are not necessarily the most important. For it may be that in the region where this large overlapping is, the value of the alteration $\mathfrak{V}$ will happen to be small. We are not interested in transitions in general, but only in transitions caused by $\mathfrak{V}$. We are asking what effect the alteration $\mathfrak{V}$ will have when it is applied, and even large overlappings of patterns in regions where $\mathfrak{V}$ is negligibly small will be irrelevant. It is obvious then that what is important is a triple mutual overlapping of a pattern Ψ_n with $\mathfrak{V}$ and with the initial Ψ_1. Only regions where a Ψ_n overlaps both $\mathcal{V}$ and the initial Ψ_1 contribute to the growth of this particular Ψ_n. The contribution made by an element of volume $dx\,dy\,dz$ is proportional to the values which these three factors have in that little volume. As in (58) the total effect will be obtained by integrating the product over all space, (62). It is scarcely necessary to point out that in this section the use of normalised values is to be understood throughout; in the growth of any Ψ_n it is the value of a_n which increases.

The above argument seems to be a reasonable method of approaching the description of events, and will be shown to follow directly from (39). But, whereas in § 1 we set out to find the value of an energy and found it in the integral (58), here we set out to find the probability of an event, and have so far found an integral (62) whose dimensions are again those of an energy. We recall that the various patterns belong to quantised levels, and that the transition from an initial state of energy W_m to a higher state of energy W_n involves the acquisition of an amount of energy $(W_n - W_m)$ from the disturbing agency. If this acquisition of a quantum involves Planck's constant, as we should expect from equation (39), we should not be surprised to find the frequency $(W_n - W_m)/h$ appearing as an oscillation frequency. In fact, writing out the integral with the overlapping patterns in full, as in (36), we obtain

$$\int \Psi_n{}^\star \mathcal{V} \Psi_m dv = \int \psi_n{}^\star \mathcal{V} \psi_m e^{2\pi i\{(W_n - W_m)/h\}t} . dv \quad ......(62).$$

In § 5 the proper use of this integral will be derived from equation (39). The derivation depends upon a certain property of ψ-functions which must be first explored.

The ψ-functions belonging to the allowed energies of any particular potential box (or any V-curve) form a family with certain properties common to all members. Taking first patterns in one dimension, consider, for example, the value of the integral

$$\int_{-\infty}^{\infty} \psi_m(x) . \psi_n{}^\star(x) . dx \quad ......(63),$$

where ψ_m and ψ_n are any two of the normalised patterns. We have already come upon one example of such an integral, namely in

(31) on p. 55; the value there turned out to be zero. As another example take the curves *a* and *b* of fig. 15 on p. 22. At any point in the right-hand half the product $\psi_1\psi_2$ is negative, while at the corresponding point in the left-hand half the product has an equal positive value. Hence in this case the value of the integral is again zero, and the same is true of any odd and even pair. Considering next curves *a* and *c* of fig. 15, we have, on multiplying the ordinates, a negative contribution from the middle part, and positive contributions from the right and left. The result is not obvious, but these do cancel each other, so that (63) is again zero. These examples are merely illustrating a general property, which is quite independent of the shape of the V-curve to which the patterns belong, so long as it is the same V for all. And it is true not only in one dimension, but for any number of dimensions; that is, the integral $\int \psi_m \psi_n^{*} dv$ taken over all space is always zero. The fact that the ψ's all satisfy the same Schroedinger equation and are zero at infinity suffices to give them this property, which is known as orthogonality.

When we are dealing with the usual composite state expressed by a sum of normalised ψ's, as in equation (29), this orthogonal property enables us to pick out the term belonging to any particular energy level that we please. Suppose that we wish to pick out from the series the term belonging to the particular energy level W_2. We multiply the whole sum by ψ_2^{*}, and integrate over all space, with the result

$$\int (a_1\psi_1 + a_2\psi_2 + a_3\psi_3 + \dots)\,\psi_2^{*} dv = a_2 \qquad \dots\dots(64).$$

For by the orthogonal property every term in the series except the second is going to be zero. And since ψ_2 is normalised, the second term yields just a_2. Use will be made of this method in what follows.

§ 5. Equation (39) tells us how the value of Ψ at any point varies with the time. But in the type of problem under discussion we are not interested in what happens at any particular point x, y, z. We are interested in how the values of $a_1, a_2, a_3, \dots$ vary with the time. And the question is then how to adapt equation

(39) for this purpose. Although the method is of general application, we may keep in mind our problem of the excitation of an atom. To avoid the necessity of handling a large number of states, we will consider only two. In the Schroedinger equation (39) write $\Psi = (a_1\Psi_1 + a_2\Psi_2)$, where a_1 and a_2 are constants:

$$\frac{ih}{2\pi}\frac{\partial}{\partial t}(a_1\Psi_1+a_2\Psi_2)+\kappa\frac{\partial^2}{\partial x^2}(a_1\Psi_1+a_2\Psi_2) - V(a_1\Psi_1+a_2\Psi_2)=0 \quad \text{......(65).}$$

If now at a certain moment the potential energy is altered from V to $(V+\mathcal{V})$, it gives us a new equation containing two extra terms, $\mathcal{V}a_1\Psi_1$ and $\mathcal{V}a_2\Psi_2$. To solve this equation, the idea is to find terms of opposite sign which will cancel out these new terms. For this purpose we suppose that the effect of the alteration in V is to make the values of a_1 and a_2 vary with the time, instead of being constants as in (64). When the first bracket is written out in full, we shall now obtain new terms in da_1/dt and da_2/dt, which were formerly zero. The whole expression is now:

$$\frac{ih}{2\pi}\left(a_1\frac{d\Psi_1}{dt}+\Psi_1\frac{da_1}{dt}+a_2\frac{d\Psi_2}{dt}+\Psi_2\frac{da_2}{dt}\right)+\kappa\frac{\partial^2}{\partial x^2}(a_1\Psi_1+a_2\Psi_2) - (V+\mathcal{V})(a_1\Psi_1+a_2\Psi_2).$$

Comparing this with (64), we see that this expression will be equal to zero, provided that

$$\frac{ih}{2\pi}\left(\Psi_1\frac{da_1}{dt}+\Psi_2\frac{da_2}{dt}\right)-\mathcal{V}(a_1\Psi_1+a_2\Psi_2)=0 \quad \text{......(66).}$$

Although for simplicity we have been using only the x-coordinate we should have obtained (66) if we had started with x, y and z.

If now the system is known to be initially in state 1, we know until time $t=0$ that $a_1=1$ and $a_2=0$. Then, as a_2 grows at the expense of a_1, we wish to know the value of a_2 at any subsequent time. A simple problem is to find the initial rate at which a_2 rises from zero. During a sufficiently short interval after $t=0$ the value of a_1 will still remain nearly equal to unity, and the value of a_2

will be negligible in comparison; that is, during this interval then we have approximately

$$\mathcal{V}(a_1\Psi_1+a_2\Psi_2)=\mathcal{V}\Psi_1 \qquad \text{......(67).}$$

Hence (66) may be written

$$\Psi_1\frac{da_1}{dt}+\Psi_2\frac{da_2}{dt}=\frac{2\pi}{ih}\mathcal{V}\Psi_1 \qquad \text{......(68).}$$

By making use of the orthogonal property described above, we may now select the term referring to state 2. Multiply both sides by $\Psi_2{}^\star$, and integrate over all space. The first term on the left-hand side becomes zero, and the second term becomes just da_2/dt:

$$\frac{da_2}{dt}=\frac{2\pi}{ih}\int\Psi_2{}^\star\mathcal{V}\Psi_1 dv \qquad \text{......(69).}$$

We see that our suggested Principle of Overlapping is vindicated; every element of volume in which Ψ_2 overlaps both $\mathcal{V}$ and Ψ_1 contributes to the rate at which the value of a_2 increases. The amount of mutual overlap measures the chance of $\mathcal{V}$ causing a transition from the state represented by Ψ_1 to that represented by Ψ_2. As mentioned in § 4, the integral has the dimensions of an energy; we see now how, in conjunction with Planck's quantum of action, it gives the probability of something happening in unit time. Integrating (69) with respect to the time

$$a_2=\frac{2\pi}{ih}\int_0^t\int\Psi_2{}^\star\mathcal{V}\Psi_1 dv\,dt \qquad \text{......(70).}$$

Although for simplicity we started with a system in its ground state, we might have started with a composite Ψ; in this case the rate of growth of a_2 would depend on transitions from each of the states initially present. The value reached by a_2 at time t may be expressed as a sum of all these contributions:

$$a_2(t)=a_{12}+a_{32}+a_{42}+\ldots \qquad \text{......(71).}$$

Any $|a_{mn}|^2$ then measures the probability that $\mathcal{V}$ causes a transition from state m to state n, and is given by

$$|a_{mn}|^2=\frac{4\pi^2}{h^2}\left|\int_0^t\int a_m\Psi_n{}^\star\mathcal{V}\Psi_m dv\,dt\right|^2 \qquad \text{......(72).}$$

The utility of equation (72) is extremely comprehensive. It enables us to calculate theoretical absorption coefficients for visible, infra-red, X-ray and γ-ray radiations, as well as the magnitudes of the photoelectric effect, and the ionisation and excitation by all kinds of moving particles.

Integrals $\int \Psi_n{}^* \mathcal{V} \Psi_m dv$ taken over all space are used so frequently that it is convenient to use an abbreviation; for this purpose the notation $\mathcal{V}_{mn}$ is used, or else $(m \mid \mathcal{V} \mid n)$. These $\mathcal{V}_{mn}$ are known as the matrix elements of $\mathcal{V}$, because when set out they form an array or matrix.

$$\begin{matrix} \mathcal{V}_{12} & \mathcal{V}_{13} & \mathcal{V}_{14} \\ & \mathcal{V}_{23} & \mathcal{V}_{24} \\ & & \mathcal{V}_{34} \end{matrix}$$

Since the volume of an atom is so small—since the region in which the atomic electron cloud is appreciable is only about 10^{-22} c.c.—problems are very common in which the intensity of the electric field E causing the disturbance $\mathcal{V}$ is uniform throughout this volume; that is, although the intensity E of the field may be changing rapidly with the time, it has at any moment the same value throughout the atom; that is to say, the V-curve for an atomic electron becomes uniformly tilted, fig. 62a or b. If we take the direction of the lines of force as the x-axis, the slope of the V-curve being ϵE, we have $\mathcal{V} = \epsilon E x$, and

$$\int \Psi_n{}^* \mathcal{V} \Psi_m dv = \epsilon E \int \Psi_n{}^* x \Psi_m dv \qquad \text{......(73).}$$

Writing the Ψ out in full,

$$\int \Psi_n{}^* x \Psi_m dv = e^{-2\pi i \nu_{mn} t} \int \psi_n{}^* x \psi_m dv \qquad \text{......(74),}$$

where $\nu_{mn} = (W_n - W_m)/h$.

Now it is clear that $\int \psi_n{}^* x \psi_m dv$ is a permanent atomic quantity, and has nothing to do with any disturbance which may or may not be applied to the atom; for brevity such an integral is usually written x_{mn}. Each of the quantities x_{mn} is a definite length, less than an atomic radius, and characteristic of the particular pair of states of the particular atom or molecule. They may be calculated once and for all, and subsequently used to estimate transitions caused by any disturbance whose field is sufficiently nearly uniform over an atomic volume. In any atom some of the quantities x_{mn} may turn out to be zero, owing to symmetry; compare (63). Thus x_{mn} will be zero for any pair of s-states of an atom.

§6. It is when we come to apply these methods to the absorption of radiation that we find in what a satisfactory way quantum mechanics joins on to classical theory. For in this problem we disregard the existence of light-quanta, and treat the incident radiation as a classical wave train. We shall take for simplicity a plane-polarised beam of frequency ν. As the wave train passes over an atom, the latter is subject to a rapidly alternating electrostatic field, whose intensity at any moment may be written $E = A\cos 2\pi\nu t$. Visible light has a wave-length of 5000 Ångström units, so that even in ultra-violet radiation the wave-length will be nearly 1000 times an atomic diameter. The intensity of the field may therefore be taken to be uniform over the atom. Taking the curve of fig. 7 as the initial V-curve for the valence electron, consider how it will be altered in different directions. In the direction of the light ray it will remain unaltered, because the field in the wave is entirely transverse. In the transverse direction the V-curve will oscillate with frequency ν between curves a and b of fig. 62. Taking x in the direction of the vector,

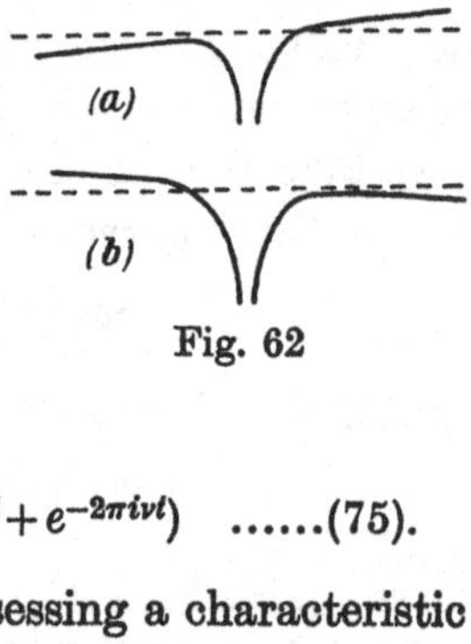

(a)

(b)

Fig. 62

$$\mathcal{V} = E\epsilon x = A\epsilon x\cos 2\pi\nu t = \frac{A\epsilon x}{2}\left(e^{2\pi i\nu t} + e^{-2\pi i\nu t}\right) \quad \text{......(75).}$$

To find how the electron in an atom possessing a characteristic frequency ν_{mn} will respond to incident light of frequency ν, we must substitute for $\mathcal{V}$ in (70). Some time before the introduction of quantum mechanics it had been realised that when an atom is initially in an excited state, incident light may have either of two alternative effects; it may cause the atom to jump to a higher excited state (absorption), or to a lower state (stimulated emission). Instead of taking an atom initially in its ground state, it will therefore be better to suppose that the initial state, of energy W_m, is an excited level, and that the energy of the final state, W_n, may be either greater or less than W_m. From (75), since x_{mn} is independent of t, we obtain

$$\int_0^t \int \Psi_n{}^{\star} \mathcal{V} \Psi_m dv\, dt = A\epsilon x_{mn} \int_0^t e^{2\pi i \nu_{mn} t} . \cos 2\pi\nu t . dt$$

$$= \tfrac{1}{2} A\epsilon x_{mn} \left[\frac{e^{2\pi i(\nu_{mn}-\nu)t}}{2\pi i(\nu_{mn}-\nu)} + \frac{e^{2\pi i(\nu_{mn}+\nu)t}}{2\pi i(\nu_{mn}+\nu)} \right]_0^t \quad \text{......(76).}$$

From this expression we can find the value of $|a_{mn}|^2$ for incident light of any wave-length we choose. Since ν_{mn} denotes $(W_n - W_m)/h$, it will be positive or negative according as the final level which we are considering is higher or lower than the initial level. When ν_{mn} is positive, the first term in (76) will obviously give resonance. The nearer the incident frequency ν is to any natural frequency ν_{mn} of the atom or molecule, the smaller will be the bracket $(\nu_{mn} - \nu)$ in the denominator, and the greater will be the response. In fact, the probability of the absorption of a quantum is negligible except when there is resonance between ν and one of the ν_{mn}, thus leading to line-absorption spectra. Whatever the intensity, the effect is strictly proportional to A^2, the square of the amplitude. The relative strengths of the various absorption lines are given by the values of $|x_{mn}|^2$; whenever this happens to be zero the line does not occur.

When we fix our attention on any possible transition down to a lower level, $(W_n - W_m)$ is negative. The bracket in the denominator of the second term in (76) can now give resonance for certain incident frequencies, for which the contribution from the first term will be negligible.

The problem of the strength of absorption lines is thus solved when we know the ψ-patterns of the system, from which to calculate x_{mn}. For the hydrogen atom and for the ions He^+ and Li^{++} the functions are known exactly, and the absorption coefficients can be calculated in absolute magnitude. Since a thermodynamic argument shows that the intensity of emission must bear a definite ratio $8\pi h\nu^3/c^3$ to the intensity of absorption, the strength of the emission lines is known as well.

For investigating the excitation of atoms, the Ψ_2 above was taken to be that of an electron bound in one of the discrete levels. But there is nothing in the equations to impose this restriction. For the final state we may insert instead any Ψ representing a free electron moving away from the atom, i.e. a Ψ belonging to the same V-curve but for a higher energy; in this case (72) will give the probability of photo-ionisation. There is of course no quantisation for free particles; hence an atom can respond to all frequencies whose $h\nu$ is greater than $\mathscr{I}$, the ionisational potential, the electron escaping with kinetic energy $\frac{1}{2}mv^2 = h\nu - \mathscr{I}$. Beyond the ultra-violet limit of the atomic line spectrum there is thus a continuous absorption spectrum, whose intensity can be calculated from (72). This treatment may be extended to electrons in the inner shells of an atom; by taking as the initial state the Ψ-pattern of a K-electron or an L-electron, the intensity of the photoelectric absorption of X-rays may be calculated. In the same way one can investigate the photoelectric effect at metal surfaces by using for Ψ_1 and Ψ_2 the patterns for an electron in the metal and for an electron escaping from the surface.

The electronic excitation of a molecule is a rather more complicated process. In the atomic problem we had only the overlapping of the two electronic patterns to consider. But for a molecule the initial and final states must each be represented by a product $\psi_{el}\psi_{vib}$. Hence

$$\psi_1\psi_2 = \psi_{el_1}\psi_{vib_1}\psi_{el_2}\psi_{vib_2} \quad \ldots\ldots(77).$$

If for any initial and final states the overlapping of ψ_{vib_1} with ψ_{vib_2} is small, this fact alone will suffice to inhibit a strong absorption of the corresponding wave-length.

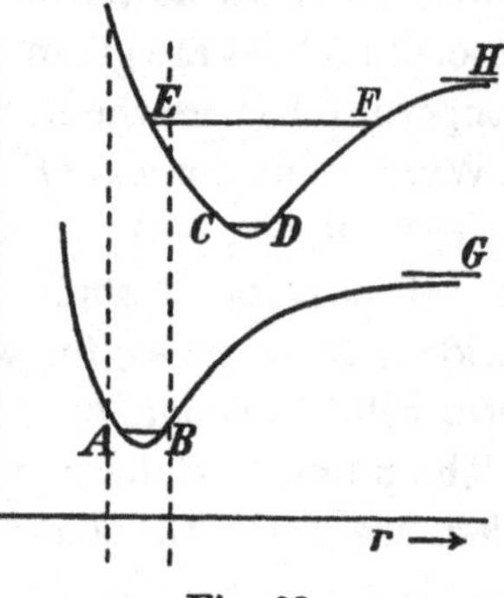

Fig. 63

The curves of fig. 63 have already been referred to on p. 89. Let CD be the lowest vibrational level of the first excited state, while AB is the lowest vibrational level of the normal state. Abscissae being the nuclear separation r, the vertical distance

GH represents the energy required for the corresponding atomic excitation when the atoms of the molecule are far apart. For the level *AB* the ψ_{vib} will be like fig. 15*a*, dying away exponentially to the right of *B* and to the left of *A*, so that its value will be small except for those values of the nuclear separation *r* lying between the vertical dotted lines which have been drawn in the diagram.

Consider now the excitation of a molecule initially in the level *AB*. The smallest energy required to raise the molecule to a higher electronic state is that given by the vertical distance between *AB* and *CD*. For this type of molecule, however, such a transition is very improbable, because the ψ_{vib} for the level *CD*, falling off exponentially from *C*, has a very small overlapping with the ψ_{vib} of the level *AB*. Any higher vibrational level near *EF* has a better overlap with *AB*. Consider next what will happen if a molecule in the level *AB* is illuminated with white light. It will absorb wave-lengths corresponding to the vertical distance *BE* much more strongly than longer wave-lengths. Further it will appear that this argument applies also to excitation of molecules by electron impact. But we will first discuss the excitation of atoms.

When an electrically charged particle passes with high speed through a gas, the atoms or molecules near its track experience an intense transient field as the particle goes by. Atoms more distant from the track experience a field which, though varying rapidly in intensity, will be nearly uniform over the atom at any moment. For an electron in the atom the *V*-curve will be uniformly tilted, as in fig. 62 *a* or *b*. This field may be relatively very weak; yet by (73) there is a probability of excitation and ionisation, however weak the field. Thus a few atoms distant from the track may be excited, the probabilities being given by the values of x_{mn}, as in the case of excitation by light. On the other hand, a large part of the ionisation and excitation, which takes place nearer the track, must be calculated from (72), taking into account that $\mathcal{V}$ is not uniform over the atom. For molecules the excitation will take place in accordance with fig. 63.

In discussing the conductivity of metals in the last chapter, it was pointed out that when an electron in the metal has been accelerated by the applied voltage, it has acquired energy, that is to say, it has made a transition to a higher level. At first sight this excitation does not appear to fall in the class of problems in which the alteration in the potential energy is transient. When a voltage is applied to two ends of a wire, the V-curve of fig. 37 is modified; but ordinarily the voltage must be kept on permanently, in order to maintain the current. This at any rate is always necessary at room temperature. If, on the other hand, we have near the absolute zero of temperature a piece of metal capable of supra-conductivity, a momentary e.m.f. is sufficient to generate a current which will persist for a long time after the e.m.f. is removed. At room temperature the thermal vibrations on their own account cause a secondary modification of the V-curve which destroys the flow of current almost as soon as it begins.

REFERENCES TO CHAPTER IX

§ 1. For Schroedinger's perturbation theory: *Annalen d. Physik*, vol. LXXX, p. 437.

§ 2. Pauling's work on valence bonds is in a series of papers in the Journal of the American Chemical Society and the Journal of Chemical Physics.

For further work on molecules: Eyring and Polanyi, *Zeit. f. Phys. Chem.* B, vol. XII, p. 279; Eyring, *J.A.C.S.* vol. LIII, p. 2537; Lennard Jones, *Proc. Phys. Soc.* vol. XLIII, p. 461; Dennison, *Rev. Modern Physics*, vol. III, p. 280; Rosen and Morse, *Phys. Rev.* vol. XLII, p. 210.

§ 5. For Dirac's perturbation theory: § 52 of his *Quantum Mechanics*.

§ 6. For the Frank-Condon Principle: Condon, *Phys. Rev.* vol. XXXII, p. 858.

For excitation and ionisation by impact: E. J. Williams, *Proc. Roy. Soc.* vol. CXXXIX, p. 163; Mott and Massey, *Atomic Collisions*.

CHAPTER X

§ 1. THE DESCRIPTION OF PHYSICAL EVENTS (*continued*)

The transitions considered in the preceding chapter were between two allowed energies W_m and W_n belonging to the same V-curve. Problems like that of fig. 44, p. 81, are equally important, but of a different type. Here the disturbance radically changes the V-curve, and calls into existence new possible motions which were formerly absent. In fig. 44 the electrons could escape from the metal A without any change in the energy W. To predict what will happen in this type of problem, we must again adapt equation (39) so as to describe the change in the Ψ-pattern. But before developing the theory it will be as well to mention other examples of this kind of physical process.

Consider a free atom *in vacuo* near a metal surface. The potential energy of the valence electron in the field of the atomic nucleus or positive core will be like fig. 7, while at the metal surface it will be like fig. 37. The V-curve along a line perpendicular to the metal surface and passing through the atom will be obtained by joining these together as in fig. 64; it comprises two potential boxes with a barrier between, whose width is the distance of the atom from the surface. For convenience suppose that the metal is at ordinary low temperature so that there is a sharp division at the critical energy between the occupied and vacant levels. The problem now is whether the valence electron from the atom will pass through the potential barrier into the metal. It cannot do so if the only vacant levels in the metal are higher than the initial level in the atom, i.e. when ϕ is less than $\mathscr{I}$. The ionisation potentials of alkali atoms, however, lie between 4 and 5 ϵ-volts; and when such an atom is incident on a metal with a sufficiently large work function, there is a high probability that it gives up its valence electron to the metal. In fact, a

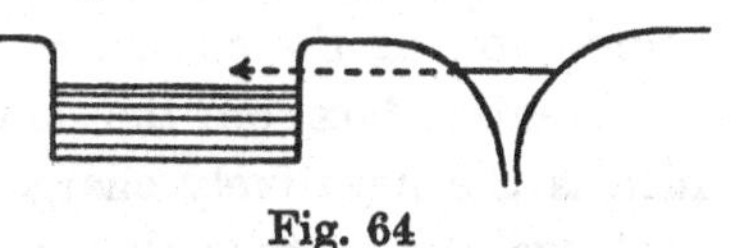

Fig. 64

standard method of producing an intense stream of ions is to direct a stream of alkali vapour at a hot filament (the high temperature prevents the incident atoms from sticking to the metal surface).

Alternatively, we may suppose that the valence level in the atom is initially vacant, i.e. we have a positive ion approaching the metal surface. An electron from the metal may now pass through the barrier to the ion, neutralising it, except when $\mathscr{I}$ is less than ϕ. This is presumably the usual process by which atomic and molecular positive ions become neutralised on approaching an electrode *in vacuo*, while negative ions lose their electrons by the reverse process above.

For a second example, suppose that we have two pieces of metal, as in the problem of fig. 44; but, instead of putting them near together, charge one to a high potential with regard to the other. The shape of the V-curve along a line passing through them will be changed from fig. 27 to fig. 65. Looking now at the electrons in the potential box on the left, which is the negatively charged metal, we notice that they are only held in the box by a potential barrier of triangular shape, which is shown on a larger scale in curve b. The width QR of this forbidden region will be less, the greater the applied voltage. And there is the possibility of electrons escaping from the metal, even at room temperature. In practice the height of the barrier is at least two volts, the value of the work function of an alkali metal; and the width QR is that distance in which the electrical potential falls by two volts; that is to say, if the intensity of the field is a million volts per centimetre, QR will be about 200 Ångström units. A triangular potential barrier will be a little more transparent to electrons than a rectangular barrier two volts high and of the same width. But from the calculation given on p. 18 we should

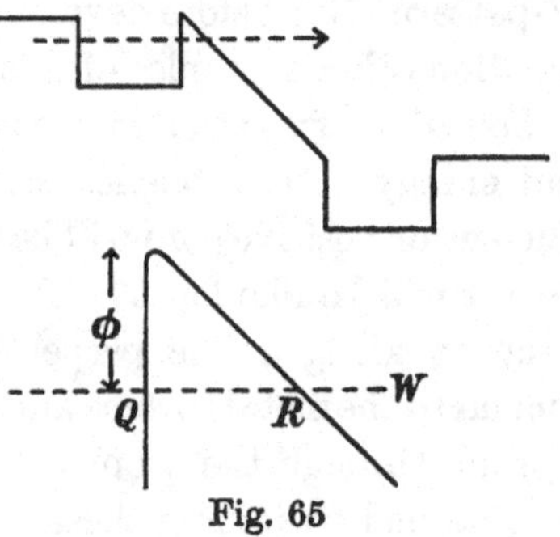

Fig. 65

hardly expect an appreciable number of excursions through the barrier until its width is well under 100 Ångström units. It is found by experiment that the escape of electrons from metals at room temperature becomes measurable for applied fields of intensity greater than 10^7 volts per centimetre.

§2. The most striking example of this leaking of particles through a forbidden region occurs in the phenomena of radioactivity, which remained a complete mystery until after the development of quantum mechanics. Stable nuclei exist for the same reason that stable atoms and molecules exist, namely, that the constituent particles move in a potential box. For radioactive elements the unanswerable question always was: Why should a nuclear particle, after remaining in the box for days or years, suddenly escape for no apparent reason?

In order to treat the expulsion of alpha particles by quantum mechanics, one must know the shape of the V-curve for the potential energy of the alpha particle in the field of the nucleus. Unlike the stable systems already discussed in this book, both the alpha particle and the nucleus bear charges of the same sign. In the study of the scattering of alpha particles by nuclei of various elements the ordinary Coulomb repulsion is in fact found, indicating a V-curve like that of fig. 46. It had, however, been recognised that at closer distances this intense repulsion must change over into an attraction, which enables the ordinary nuclei to exist in spite of their large positive charge. The required V-curve will be like fig. 46, with some kind of potential box in the middle. From a V-curve like fig. 66 we can see how both stable and unstable nuclei may exist. A particle inside the nucleus, whose energy W is to be represented by a horizontal line lying below the axis OX, cannot escape into surrounding space. With a higher occupied level, on the other hand, such as AB, if the ideas under discussion are correct, it should be possible to represent the state by a Ψ-pattern which would leak out steadily, even if it took centuries to do so. In this way one could account for the existence of the radioactive elements.

§3. Before linking up this type of problem with those of

Chapter IX, a more direct approach would seem to be a discussion of a V-curve like that of fig. 66b, where the straight lines ST and PO are supposed to extend to infinity in either direction. A potential box PQ is separated from empty space by a barrier QR. In looking for acceptable ψ-patterns, we find at once that for this potential box there are no discrete quantised levels. In the region ST there will be a uniform sine curve, extending to infinity; while in PQ will be a portion of the same sine curve of different amplitude. In fig. 27 the ψ-curve had to fall exponentially to zero on the right as well as on the left; remove this restriction, and at once there are acceptable ψ-patterns for all values of W. These may be divided into ψ-curves for which the amplitude of ψ is (a) small in PQ compared with the amplitude outside in ST, and (b) large in PQ compared with that outside. The ψ-curve under the barrier QR being given by (4) the relative amplitude outside depends on whether D is greater or less than Ce^{-kd}, where d is the width of the barrier QR. All kinetic energies from zero upwards being allowed, for the lowest energies the amplitude of ψ inside PQ is very small, because the wave-lengths are much too long to fit approximately into PQ. Looking at higher energies, we come to a range whose ψ will fit into PQ with small values of D; this narrow range occurs round what would be the lowest allowed level of PQ if the barrier QR were infinitely wide. Every ψ-pattern in the range will vibrate with its own frequency W/h; and, if a wave packet is formed from them to represent a particle initially in PQ, they will get out of phase, so that the amplitude in PQ decays with the time.

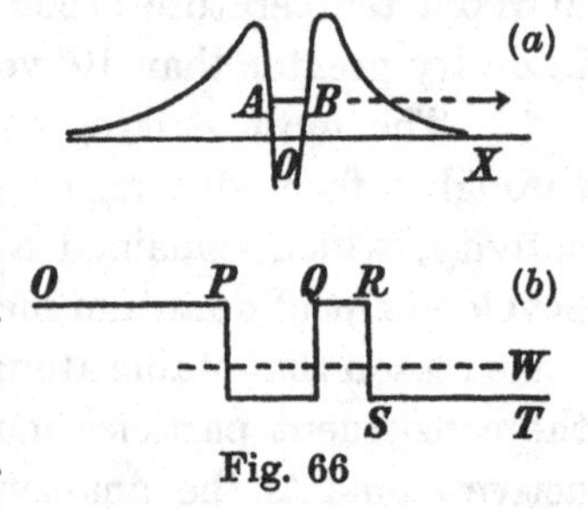

Fig. 66

§4. Looking now at the physical problems in which this leaking through a potential barrier occurs, we see that such a leak cannot have been going on for all time. In every case it is a temporary phenomenon due to the fact that the present V-curve is different from what it was during some previous period. In the problem of fig. 44 the two pieces of metal had been far apart, and an appreciable transfer of electrons only started when they were brought nearer together. In fig. 64 the barrier between the atom and the metal only becomes narrow when the atom approaches the surface. In fig. 65 electrons only begin to escape when the V-curve is made sufficiently steep by a large applied voltage. And in fig. 66 the leak from any radioactive nucleus only begins at the moment when the parent atom in the series disintegrates. For example, Radium F (polonium) emits alpha particles while its parent substance Radium E does not (at any rate not in observable quantity). The disintegration of Radium E evidently leaves behind a situation like fig. 66*a*, say at time $t=0$, so that the steady leak of an alpha particle starts from this moment.

This point of view suggests that equation (39) may be adapted for describing the physical process by a method similar to that of Chapter IX, namely, by modifying the initial potential energy V by a $\mathcal{V}$-curve, to give $(V+\mathcal{V})$. For example, let curve *a* of fig. 67 be the initial V-curve with an isolated potential box PQ, which will have discrete levels. If we start with a particle in the lowest level of PQ, we have an initial state without any ambiguity; the Ψ_1 falls off exponentially outside the box. Now take

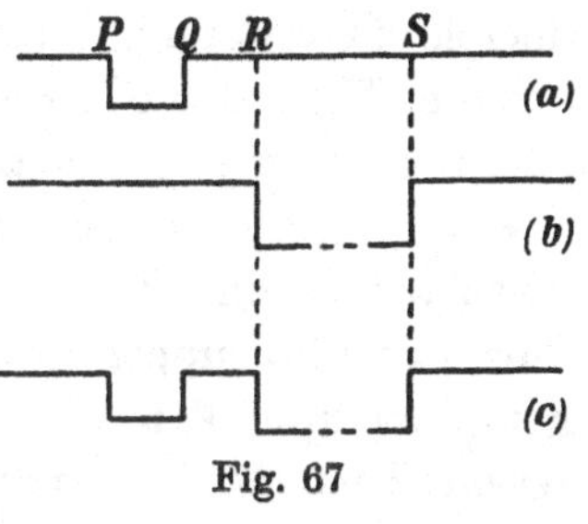

Fig. 67

curve *b* as the $\mathcal{V}$-curve, the disturbance being zero except between R and S. On adding the ordinates, the sum $(V+\mathcal{V})$ gives curve *c* with two potential boxes. It will be convenient to suppose that the box RS is much wider, compared with PQ, than is shown in the diagram, for then the allowed levels will be very close together, and there are sure to be a few levels of nearly the same energy as the initial level in PQ. Suppose then that at time $t=0$ this box RS comes into existence, with a set of levels all of which are empty. The problem is to find the rate at which a Ψ-pattern springs up in the box RS owing to the fact that RS overlaps the exponential tail of the initial Ψ_1. As in Chapter IX, let us fix attention first on one only of the possible final states, which, for comparison with (64), we may denote by Ψ_2. Whereas in Chapter IX both Ψ_2 and Ψ_1 belonged to the V-curve, here Ψ_2 is a characteristic pattern of the $\mathcal{V}$-curve, and satisfies the equation

$$\frac{ih}{2\pi}\frac{\partial}{\partial t}a_2\Psi_2+\kappa\frac{\partial^2}{\partial x^2}a_2\Psi_2-\mathcal{V}a_2\Psi_2=0 \quad \text{......(78).}$$

Writing down the similar equation with a_1, Ψ_1 and V, and adding, an equation identical with (65) is obtained except that the term $\mathcal{V}a_2\Psi_2$ occurs in place of $Va_2\Psi_2$. To obtain a solution of the equation with $(V+\mathcal{V})$ we follow the same argument as before, and arrive at an equation identical with (66) except that $Va_2\Psi_2$ occurs in place of $\mathcal{V}a_2\Psi_2$; this is because it is the box PQ which is foreign to Ψ_2, while the box RS is foreign to Ψ_1. But during a short interval of time a_2 remains small compared with a_1, so that (68) is again true, and brings us again to equation (69). In spite of the difference between this problem and the previous type, the Principle of Overlapping may be applied in exactly the same form as before. The value of da_2/dt now gives the rate at which the pattern Ψ_2 springs up in the empty box RS, and hence the probability of the particle passing through the barrier into the particular level on which we are fixing attention.

For simplicity suppose that the potential energy, after changing rapidly from V to $(V+\mathcal{V})$, remains constant for a certain interval. Then, $\mathcal{V}$ being independent of the time, (70) may be written

$$a_2=-\frac{2\pi}{ih}\int\psi_2{}^*\mathcal{V}\psi_1 dv\int_0^t e^{2\pi i\{(W_2-W_1)/h\}t}.dt \quad \text{......(79).}$$

The probability of a transition from any initial level to any final level will again be given by the square of a matrix element. In order to deal with problems like those of figs. 65 and 66, in which a particle escapes into free space, it will be necessary later to suppose that the box RS is indefinitely long; then the set of allowed levels of RS will close up into a continuous range of allowed energy. In any case let W_2 be some energy in the neighbourhood of the initial level W_1 of the box PQ. The factor $e^{2\pi i\{(W_2-W_1)/h\}t}$ in (79) oscillates between positive and negative values with a period $\tau=h/(W_2-W_1)$. And if it has changed sign several times within the short interval considered, the positive and negative portions of the integral will nearly cancel each other out, leading to a negligibly small value of $|a_2|$. In order that the factor may not have changed sign several times during the interval, the value of τ must be large, which occurs only when (W_2-W_1) is small. That is to say, the only Ψ-patterns which spring up in the box RS are those which have almost exactly the same energy as the initial level W_1. The slight uncertainty in the energy, depending upon the length of the interval, is to be expected from (27).

Expression (79) applies to a potential barrier of any shape, provided that it is not too small. We should expect the probability of a transition to be in agreement with the usual estimate of excursions into a forbidden region for a particle of any mass. That is, in fig. 67, for example, it should depend upon the amount by which the exponential tail of the pattern ψ_1 extends beyond the point R. We see that this will be so in (79). For, since $\mathcal{V}$ is zero except in the region RS, the value of the integral will depend upon the amount to which the exponential tail of ψ_1 extends into this region. If the distance QR is d, the probability of a transition will be proportional to e^{-2kd}. Potential barriers are usually too broad to allow the passage of particles having a greater mass than a proton. In the nucleus, however, we find barriers of width less than 10^{-12} cm., which are transparent to alpha particles. All the methods of treating the problem of fig. 66*b* show that the probability of escape per second may be taken to be of magnitude νe^{-2kd}, where ν is the frequency of the to and fro motion of the particle in the box. It is useful to remember that for a potential barrier of height one ϵ-volt, and breadth one Ångström unit, the value of e^{-2kd} for an electron is almost exactly e^{-1}, or $1/e$ (compare p. 19). For an electron in an atom the value of ν may be taken as roughly 10^{15} per second, and thus the probability of escape through any barrier may be roughly estimated.

In the V-curve and $\mathcal{V}$-curve of fig. 67 we were concerned with the potential energy of a single particle. We could, on the other hand, start with a V-surface for a pair of particles; modifying this by means of a $\mathcal{V}$-surface, we could introduce new regions where V will be less than the initial W. We can then find the rate at which the Ψ-pattern, representing the pair of particles, leaks into these new regions.

§5. In the problems of excitation by electron impact, etc., discussed in the last chapter, the atom or molecule was raised to a state of higher energy W_2. For this it was essential that the disturbance $\mathcal{V}$ should vary with the time. For, as we have just seen, when $\mathcal{V}$ is constant, equation (70) becomes (79), and provides transitions only between states of the same energy. On looking at the process more closely, however, we recall that when a system is raised to a higher level, the energy must have come from somewhere. Energy is conserved; the transition is really from a certain state to another state of the same energy. In the case of excitation by impact, the impinging particle loses as much energy as the atom gains. In fact, transitions are always between states of equal energy when the whole system is taken into account. Hence the method of this chapter is for most problems more fundamental than that of Chapter IX. The transitions through potential barriers which we have been considering must be merely one class among many for which equation (79) may be used.

With reference to figs. 5, 7, etc., it has been pointed out that a W-line lying wholly above the V-curve represents the energy of a free particle. At first sight one would expect that a W-plane lying well above a V-surface, such as that of fig. 47, would represent a pair of free particles. But W is the sum of the energies of the two particles; and one particle may be a free particle with nearly all the energy, while the other is bound in the lowest level of the potential box. For such a value of W there are evidently several states of the same energy. For, alternatively, the free particle may have rather less kinetic energy, while the other particle is in the first (or higher) excited level of the potential box. These two states will be represented by distinct Ψ-patterns belonging to the same energy. If a transition takes place between the first state and the second, the atom becomes excited, while the free particle loses an equivalent amount of kinetic energy. This is, in fact, the familiar process of excitation by impact. Naturally the transition only occurs when the particles collide. If the energy of the interaction between the atomic electron and the free particle is denoted by $\mathcal{V}$, we may look upon $\mathcal{V}$ as causing the excitation. Whereas in fig. 62 $\mathcal{V}$ was a $\mathcal{V}$-curve varying with the time, here it is a $\mathcal{V}$-surface independent of the time (a diagonal ridge or the equivalent in six dimensions). When $\mathcal{V}$ causes a transition from Ψ_1 (belonging to energy W_1) to Ψ_2 (belonging to energy W_2), use of equation (79) ensures that the reduction in the energy of the impinging particle will not differ by a measurable amount from the excitation energy of the atom.

REFERENCES TO CHAPTER X

§ 1. For the problem of fig. 64: Massey, *Proc. Camb. Phil. Soc.* vol. XXVI, p. 386.

For the problem of fig. 65: Fowler and Nordheim, *Proc. Roy. Soc.* vol. CXIX, p. 173; Stern, Gossling, and Fowler, *ibid.* vol. CXXIV, p. 699.

§ 3. For the leaking out from a potential box: Fowler and Wilson, *Proc. Roy. Soc.* vol. CXXIV, p. 493.

For the escape of alpha particles from the nucleus: Gurney and Condon, *Phys. Rev.* vol. XXXIII, p. 127; Gamow, *Atomic Nuclei and Radio-activity.*

§ 4. With fig. 67 compare Born, *Zeit. f. Physik*, vol. LVIII, p. 306.

§ 5. Mott and Massey, *Atomic Collisions.*

APPENDIX

NOTE 1. CENTRAL FORCE PROBLEM

In spherical polar co-ordinates equation (9), p. 25, becomes

$$\frac{1}{r^2}\frac{\partial}{\partial r}\left(r^2\frac{\partial\psi}{\partial r}\right)+\frac{1}{r^2\sin^2\theta}\frac{\partial^2\psi}{\partial\phi^2}+\frac{1}{r^2\sin\theta}\frac{\partial}{\partial\theta}\left(\sin\theta\frac{\partial\psi}{\partial\theta}\right)+\frac{8\pi^2\mu}{h^2}(W-V)\psi=0 \quad \text{......(80).}$$

Compare Jeans, *Electricity and Magnetism*, Chapter VIII. It is well known that for this type of equation a solution can be obtained as a product of two factors, of which one is a function of r only, and the other a function of θ and ϕ only. Write

$$\psi=R(r)\,.\,S(\theta,\phi),$$

substitute in (80), and multiply through by r^2/RS. We obtain

$$\frac{1}{R}\frac{d}{dr}\left(r^2\frac{dR}{dr}\right)+\left[\frac{1}{S\sin^2\theta}\frac{\partial^2 S}{\partial\phi^2}+\frac{1}{S\sin\theta}\frac{\partial}{\partial\theta}\left(\sin\theta\frac{\partial S}{\partial\theta}\right)\right]+\frac{8\pi^2\mu r^2}{h^2}(W-V)=0 \quad \text{......(81).}$$

The large bracket is independent of r, and must be constant; it has the dimensions of a pure number; set it equal to $-G$:

$$\frac{1}{S\sin^2\theta}\frac{\partial^2 S}{\partial\phi^2}+\frac{1}{S\sin\theta}\frac{\partial}{\partial\theta}\left(\sin\theta\frac{\partial S}{\partial\theta}\right)+G=0 \quad \text{......(82).}$$

Repeating the process, let $S(\theta,\phi)$ be the product of two factors, of which one is a function of θ only, and the other of ϕ only. Writing

$$S(\theta,\phi)=\Theta(\theta)\,.\,\Phi(\phi),$$

we obtain

$$\frac{1}{\Phi\sin^2\theta}\frac{d^2\Phi}{d\phi^2}+\frac{1}{\Theta\sin\theta}\frac{d}{d\theta}\left(\sin\theta\frac{d\Theta}{d\theta}\right)+G=0 \quad \text{......(83).}$$

On multiplying through by $\sin^2\theta$ the first term becomes independent of θ and must be constant; for convenience let this constant be written $-m^2$; then we have the simple equation

$$\frac{d^2\Phi}{d\phi^2}+m^2\Phi=0 \qquad \text{......(84).}$$

Equation (83) now becomes

$$\frac{1}{\sin\theta}\frac{d}{d\theta}\left(\sin\frac{d\Theta}{d\theta}\right)+\left(G-\frac{m^2}{\sin^2\theta}\right)\Theta=0 \qquad \text{......(85).}$$

We notice that when $\theta=0$, $1/\sin\theta$ becomes infinite, and there are only certain values of G for which the solution remains finite. These are known to be the integral values 0, 2, 6, 12, etc., given by $G=l(l+1)$, where l is an integer. And there is the further restriction that m must not be greater than l nor less than $-l$.

Setting the bracket in (81) equal to $-G$, we have

$$\frac{1}{R}\frac{d}{dr}\left(r^2\frac{dR}{dr}\right)+\frac{8\pi^2\mu r^2}{h^2}(W-V)-l(l+1)=0 \text{ ......(86).}$$

This may be simplified by introducing a new function

$$F(r)=r\,.\,R(r);$$

then (86) takes the form

$$\frac{d^2F}{dr^2}+\frac{8\pi^2\mu}{h^2}\left(W-V-\frac{l(l+1)h^2}{8\pi^2\mu r^2}\right)F=0 \qquad \text{......(87).}$$

NOTE 2

From the three series:

$$e^x=1+x+\frac{x^2}{2!}+\frac{x^3}{3!}+\frac{x^4}{4!}+\dots,$$

$$\cos x=1-\frac{x^2}{2!}+\frac{x^4}{4!}-\dots,$$

$$\sin x=x-\frac{x^3}{3!}+\frac{x^5}{5!}-\dots,$$

it follows at once that

$$\left.\begin{aligned}e^{i\theta}&=\cos\theta+i\sin\theta\\ e^{-i\theta}&=\cos\theta-i\sin\theta\end{aligned}\right\} \qquad \text{......(88),}$$

$$\cos\theta=\tfrac{1}{2}(e^{i\theta}+e^{-i\theta}),\quad i\sin\theta=\tfrac{1}{2}(e^{i\theta}-e^{-i\theta}).$$

If $z=a+ib$, and $z^\star=a-ib$, where a and b are any two real quantities, z and z^* are said to be conjugate to each other. The positive square root of the product zz^* is known as the modulus of z, and is written $|z|$. Here $|z|^2 = a^2 + b^2$.

From (88) it is clear that $e^{i\theta}$ and $e^{-i\theta}$ are complex conjugates; the modulus in this case is unity.

NOTE 3

It appears from (33) that differentiating Ψ with respect to the time is equivalent to multiplying it by $\frac{2\pi}{ih} W$; or

$$W = \frac{ih}{2\pi} \frac{\partial}{\partial t} \qquad \text{......(89).}$$

This relation between W, h and t shows a resemblance to (27):

$$\Delta W \sim \frac{h}{\Delta t}.$$

If this resemblance is significant, we may expect an analogous relation for (26):

$$\Delta p_x \sim \frac{h}{\Delta x}.$$

By analogy with (89) we should expect

$$p_x = \frac{ih}{2\pi} \frac{\partial}{\partial x},$$

with similar expressions for p_y and p_z. These expressions are, in fact, important for throwing light on the meaning of the Schroedinger equation.

We begin with the fact that the total energy W is the sum of the potential energy V and the kinetic energy $\frac{1}{2}mv^2$. The latter may be written $(mv)^2/2m$, or $p^2/2m$:

$$W = V + p^2/2m,$$

$$p_x{}^2 + p_y{}^2 + p_z{}^2 - 2m(W - V) = 0.$$

Substituting for p_x, p_y, p_z, we have

$$\frac{\partial^2}{\partial x^2} + \frac{\partial^2}{\partial y^2} + \frac{\partial^2}{\partial z^2} + \frac{8\pi^2 m}{h^2}(W - V) = 0 \qquad \text{......(90).}$$

On inserting ψ we have exactly (9), the Schroedinger equation.

NOTE 4. TWO PARTICLES

Let a particle of mass m_1 move along a straight line, its distance from the origin being x_1, and the potential energy along this line being $V_1(x_1)$. Let another particle of mass m_2 move along a straight line, the distance from its origin being x_2, and its potential energy $V_2(x_2)$. First suppose that these particles neither attract nor repel one another; then their ψ-functions are obtained from the equations

$$\frac{1}{m_1}\frac{d^2\psi_1}{dx_1^2}+\frac{1}{g}(W_1-V_1)\psi_1=0,\quad \frac{1}{m_2}\frac{d^2\psi_2}{dx_2^2}+\frac{1}{g}(W_2-V_2)\psi_2=0,$$

where $g=h^2/8\pi^2$.

Multiply the first of these equations by ψ_2 and the second by ψ_1, and add them together:

$$\frac{\psi_2}{m_1}\frac{d^2\psi_1}{dx_1^2}+\frac{\psi_1}{m_2}\frac{d^2\psi_2}{dx_2^2}+\frac{1}{g}(W_1+W_2-V_1-V_2)\psi_1\psi_2=0 \quad \text{...(91)}.$$

Since ψ_2 is not a function of x_1, $\psi_2\frac{d^2\psi_1}{dx_1^2}$ is equal to $\frac{d^2}{dx_1^2}\psi_1\psi_2$, and the same for the next term.

Hence writing

$$W_1+W_2=W,\quad V_1+V_2=V(x_1, x_2),$$

(91) takes the form

$$\frac{1}{m_1}\frac{\partial^2\psi}{\partial x_1^2}+\frac{1}{m_2}\frac{\partial^2\psi}{\partial x_2^2}+\frac{1}{g}(W-V)\psi=0 \quad \text{......(92)},$$

where $\psi=\psi_1\psi_2$. We have proved that for any pair of particles that do not interact the proper form of ψ is a product; which agrees with the general theorem for two independent observable quantities, (55) and (56).

So long as V is merely the sum of V_1 and V_2, (92) merely embodies the two equations from which we started. But we go on now to make the assumption that (92) is the right equation for giving the ψ-patterns for a pair of particles which do attract or repel one another.

NOTE 5

Consider two interacting particles in otherwise field-free space. The potential energy depends only on their distance apart, and not on the position of their centre of gravity. Let them move along the same straight line; and, taking some point on this line as their common origin, let their distances from the origin be x_1 and x_2 respectively. Let the distance of their centre of gravity from the origin be denoted by X, so that

$$(m_1+m_2)X = m_1x_1+m_2x_2.$$

Let ξ denote their distance apart:

$$\xi = x_2 - x_1.$$

The potential energy depends on ξ only, and may be written $V = V_0 + V(\xi)$, where V_0 is a constant depending on the zero of energy from which we choose to measure V. Now

$$\frac{\partial}{\partial x_1} = \frac{\partial X}{\partial x_1}\frac{\partial}{\partial X} + \frac{\partial \xi}{\partial x_1}\frac{\partial}{\partial \xi} = \frac{m_1}{m_1+m_2}\frac{\partial}{\partial X} - \frac{\partial}{\partial \xi}.$$

Hence

$$\frac{\partial^2}{\partial {x_1}^2} = \left\{\frac{m_1}{m_1+m_2}\frac{\partial}{\partial X} - \frac{\partial}{\partial \xi}\right\}^2,$$

$$\frac{\partial^2}{\partial {x_2}^2} = \left\{\frac{m_2}{m_1+m_2}\frac{\partial}{\partial X} + \frac{\partial}{\partial \xi}\right\}^2.$$

Substituting in (92) and using (37) we obtain

$$\frac{ih}{2\pi}\frac{\partial \Psi}{\partial t} + V\Psi - \frac{g}{M}\frac{\partial^2 \Psi}{\partial X^2} - \frac{g}{\mu}\frac{\partial^2 \Psi}{\partial \xi^2} = 0 \qquad \text{......(93)},$$

where M is the total mass, m_1+m_2, and μ is the "reduced mass", $m_1m_2/(m_1+m_2)$.

Equation (93) may be separated into two equations by the same method as was used for splitting up equation (80). That is, we try the effect of supposing that Ψ is a product of two factors, one of which, Ψ_a, is a function of the position of the centre of gravity of the particles only, and the other, Ψ_b, a function of their distance apart only:

$$\Psi = \Psi_a(X) \,.\, \Psi_b(\xi) \qquad \text{......(94).}$$

Substituting (94) in (93), and dividing through by Ψ_a, we obtain

$$\frac{ih}{2\pi}\frac{\partial \Psi_b}{\partial t}+V\Psi_b-\frac{g}{\mu}\frac{\partial^2 \Psi_b}{\partial \xi^2}+\left[\frac{ih}{2\pi\Psi_a}\frac{\partial \Psi_a}{\partial t}-\frac{g}{M\Psi_a}\frac{\partial^2 \Psi_a}{\partial X^2}\right]\Psi_b=0.$$

The bracket in this equation is a function of X only, and the other terms, including $V\Psi_b$, are functions of ξ only. The bracket must be constant. It has the dimensions of an energy. If we set it equal to $-V_0$, we have the following pair of simple equations:

$$\frac{ih}{2\pi}\frac{\partial \Psi_a}{\partial t}+V_0\Psi_a-\frac{g}{M}\frac{\partial^2 \Psi_a}{\partial X^2}=0 \qquad \text{......(95)},$$

$$\frac{ih}{2\pi}\frac{\partial \Psi_b}{\partial t}+V(\xi)\Psi_b-\frac{g}{\mu}\frac{\partial^2 \Psi_b}{\partial \xi^2}=0 \qquad \text{......(96)}.$$

Comparing (95) with (39), we see that it is simply the equation for a particle of mass m_1+m_2 in field-free space.

Although in Notes 4 and 5 only one dimension has been taken, it is clear that similar results will be obtained using x, y, and z.

Note 6

To understand under what circumstances a sum or difference, $\psi_A \pm \psi_B$, gives a good approximation to the correct patterns, as in fig. 50, we may carry out an argument like that used for equation (78). Taking a V-curve with a single potential box PQ, and another with a single potential box RS, as in fig. 67, we shall call them V_A and V_B, instead of V and $\mathcal{V}$. The equation satisfied by ψ_A is

$$\frac{d^2\psi_A}{dx^2}+kW\psi_A-kV_A\psi_A=0 \qquad \text{......(97)}.$$

Writing down the equation containing V_B, satisfied by ψ_B, first add it to (97), and secondly subtract it from (97), and we obtain

$$\frac{d^2}{dx^2}(\psi_A \pm \psi_B)+kW(\psi_A \pm \psi_B)-k(V_A\psi_A \pm V_B\psi_B)=0 \ \text{......(98)}.$$

Since we are looking for patterns appropriate to the potential energy (V_A+V_B), with two boxes, we may compare (98) with the expression

$$\frac{d^2}{dx^2}(\psi_A \pm \psi_B)+kW(\psi_A \pm \psi_B)-k(V_A+V_B)(\psi_A \pm \psi_B) \quad \text{...(99)},$$

which would be equal to zero if $(\psi_A \pm \psi_B)$ were the desired patterns. We see that (99) differs from the foregoing by the presence of two extra terms, $V_B\psi_A$ and $V_A\psi_B$, which are missing from (98). If in any particular case these extra terms happen to be negligibly small, then the correct patterns will not differ appreciably from $(\psi_A+\psi_B)$ and $(\psi_A-\psi_B)$. Returning to figs. 50 and 67, we see that so long as the boxes are far apart these are just the conditions which are fulfilled there. By definition V_A only differs from zero between P and Q, and here ψ_B is very small; similarly V_B only differs from zero between R and S, and here ψ_A is small. Hence the missing products are unimportant, in accordance with the Principle of Overlapping; and both $(\psi_A+\psi_B)$ and $(\psi_A-\psi_B)$ are good approximations to the correct patterns. If the boxes are brought nearer together, we see why the approximation becomes progressively less good.

Note 7. Derivation of (58)

When V is altered to $(V+\mathscr{V})$, let each initial level W_n be shifted to a slightly different energy $(W_n+\mathscr{W}_n)$, while the corresponding ψ_n is altered to a slightly different pattern $(\psi_n+\phi_n)$. We wish first to show that the value of each $\mathscr{W}_n$ is given by the overlapping integral (58), and secondly to modify each ψ_n by adding the most suitable small portions of the other initial patterns. We will fix attention on one particular level, say the level W_2, as in (61). To avoid having to deal with the whole series of levels at once, we will first modify ψ_2 by the addition of portions of ψ_1 and ψ_3 only; that is to say,

$$\psi_2'=\psi_2+\phi_2,$$

where

$$\phi_2=a_1\psi_1+a_3\psi_3 \quad \text{......(100)}.$$

Each original pattern satisfies equation (9), which may be written in the abbreviated form

$$\left(\frac{1}{\kappa}\nabla^2 - V\right)\psi_n = -W_n\psi_n \qquad \text{......(101)},$$

while each modified pattern satisfies the equation

$$\frac{1}{\kappa}\nabla^2(\psi_n+\phi_n) - (V+\mathcal{V})(\psi_n+\phi_n) + (W_n+\mathcal{W}_n)(\psi_n+\phi_n) = 0 \qquad \text{......(102)}.$$

Since the alterations $\mathcal{W}$ and ϕ are to be small, we may neglect the products $\mathcal{W}\phi$ and $\mathcal{V}\phi$. Subtracting (101) from (102) and writing $n=2$, we obtain the equation satisfied by ϕ_2:

$$\left(\frac{1}{\kappa}\nabla^2 - V\right)\phi_2 + W_2\phi_2 + \mathcal{W}_2\psi_2 = \mathcal{V}\psi_2 \qquad \text{......(103)}.$$

Substituting (100) in (103), we have

$$a_1\left(\frac{1}{\kappa}\nabla^2 - V\right)\psi_1 + a_3\left(\frac{1}{\kappa}\nabla^2 - V\right)\psi_3 + W_2a_1\psi_1 + W_2a_3\psi_3 + \mathcal{W}_2\psi_2 = \mathcal{V}\psi_2 \qquad \text{......(104)}.$$

We see now that by (101) the first term in (104) is equal to $-a_1W_1\psi_1$, while the second term is equal to $-a_3W_3\psi_3$. Hence (104) becomes

$$a_1(W_2-W_1)\psi_1 + a_3(W_3-W_1)\psi_3 + \mathcal{W}_2\psi_2 = \mathcal{V}\psi_2 \qquad \text{...(105)}.$$

Consider now what will happen if we multiply every term in this equation by ψ_2^* and integrate every term over all space as was done in (64). Since ψ_1, ψ_2 and ψ_3 all belong to the same system, they are orthogonal, and the first two terms when integrated will become zero. The third term becomes just $\mathcal{W}_2$. Hence

$$W_2' - W_2 = \mathcal{W}_2 = \int \mathcal{V}\,|\psi_2|^2\,dv,$$

which is just the overlapping integral. If in (100) we had had an infinite number of terms, instead of only two specimen terms, the same result would clearly have been reached.

Consider next what will happen if we multiply every term in (105) by ψ_1^*, and again integrate over all space. The second and third terms now become zero, leaving

$$a_1(W_2 - W_1)\int\psi_1\psi_1^* dv = \int\psi_1^* \mathcal{V}\psi_2 dv.$$

Hence the best value of a_1 to insert in (61) is obtained as a ratio between two energies

$$a_1 = \frac{1}{W_2 - W_1}\int\psi_1^* \mathcal{V}\psi_2 dv \qquad \text{......(106).}$$

The value a_3 is obtained by multiplying (105) by ψ_3^*; and again the value does not depend on how many terms (105) contains. The value of each a_n is found in the same way, so long as the levels are not degenerate.

ELECTRON CONFIGURATIONS AND FIRST IONISATION POTENTIALS

		K	*L*		*M*			*N*			$\mathcal{J}$
		1*s*	2*s*	2*p*	3*s*	3*p*	3*d*	4*s*	4*p*	4*d*	ε-volts
H	1	1	.	.	.	.	.	.	.	.	13·53
He	2	2	.	.	.	.	.	.	.	.	24·47
Li	3	2	1	.	.	.	.	.	.	.	5·37
Be	4	2	2	.	.	.	.	.	.	.	9·28
B	5	2	2	1	.	.	.	.	.	.	8·28
C	6	2	2	2	.	.	.	.	.	.	11·22
N	7	2	2	3	.	.	.	.	.	.	14·48
O	8	2	2	4	.	.	.	.	.	.	13·55
F	9	2	2	5	.	.	.	.	.	.	18·6
Ne	10	2	2	6	.	.	.	.	.	.	21·47
Na	11	2	2	6	1	.	.	.	.	.	5·12
Mg	12				2	.	.	.	.	.	7·61
Al	13				2	1	.	.	.	.	5·96
Si	14		10		2	2	.	.	.	.	8·12
P	15		Neon core		2	3	.	.	.	.	10·3
S	16				2	4	.	.	.	.	10·3
Cl	17				2	5	.	.	.	.	12·96
A	18				2	6	.	.	.	.	15·68
K	19	2	2	6	2	6	.	1	.	.	4·32
Ca	20						.	2	.	.	6·09
Sc	21						1	2	.	.	6·7
Ti	22						2	2	.	.	6·81
V	23						3	2	.	.	6·76
Cr	24			18			5	1	.	.	6·74
Mn	25			Argon core			5	2	.	.	7·41
Fe	26						6	2	.	.	7·83
Co	27						7	2	.	.	8·5
Ni	28						8	2	.	.	7·67
Cu	29	2	2	6	2	6	10	1	.	.	7·68
Zn	30							2	.	.	9·36
Ga	31				28			2	1	.	5·97
Ge	32				Copper core			2	2	.	8·09

SUBJECT INDEX

NAME INDEX

Printed in the United States
By Bookmasters